Unit 5
Senses and Survival

To obtain permission(s) or for inquiries, submit a request to:

Discovery Education, Inc.
4350 Congress Street, Suite 700
Charlotte, NC 28209
800-323-9084
Education_Info@DiscoveryEd.com

ISBN 13: 978-1-68220-802-1

Printed in the United States of America.

5 6 7 8 9 10 CWM 26 25 24 23   B

Acknowledgments

Acknowledgment is given to photographers, artists, and agents for permission to feature their copyrighted material.

Cover and inside cover art: Shahar Shabtai / Shutterstock.com

© Discovery Education | www.discoveryeducation.com

Table of Contents

Concept 5.3 Light and Sight

Concept 5.4 Communication and Information Transfer

Unit Wrap-Up

Grade 4 Resources

Discovery
EDUCATION

Dear Parent/Guardian,

This year, your student will be using Science Techbook™, a comprehensive science program developed by the educators and designers at Discovery Education and written to the Next Generation Science Standards (NGSS). The NGSS expect students to act and think like scientists and engineers, to ask questions about the world around them, and to solve real-world problems through the application of critical thinking across the domains of science (Life Science, Earth and Space Science, Physical Science).

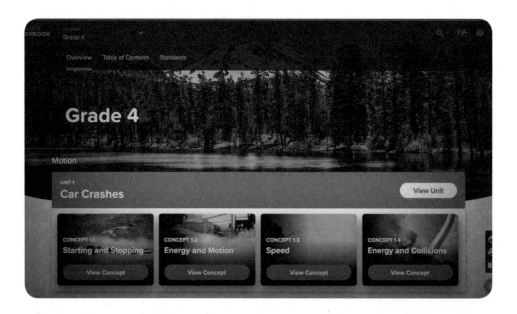

Science Techbook is an innovative program that helps your student master key scientific concepts. Students engage with interactive science materials to analyze and interpret data, think critically, solve problems, and make connections across science disciplines. Science Techbook includes dynamic content, videos, digital tools, Hands-On Activities and labs, and game-like activities that inspire and motivate scientific learning and curiosity.

You and your child can access the resource by signing in to www.discoveryeducation.com. You can view your child's progress in the course by selecting the Assignment button.

Discovery EDUCATION

Science Techbook is divided into units, and each unit is divided into concepts. Each concept has three sections: Wonder, Learn, and Share.

Units and Concepts Students begin to consider the connections across fields of science to understand, analyze, and describe real-world phenomena.

Wonder Students activate their prior knowledge of a concept's essential ideas and begin making connections to a real-world phenomenon and the **Can You Explain?** question.

Learn Students dive deeper into how real-world science phenomenon works through critical reading of the Core Interactive Text. Students also build their learning through Hands-On Activities and interactives focused on the learning goals.

Share Students share their learning with their teacher and classmates using evidence they have gathered and analyzed during Learn. Students connect their learning with STEM careers and problem-solving skills.

Within this Student Edition, you'll find QR codes and quick codes that take you and your student to a corresponding section of Science Techbook online. To use the QR codes, you'll need to download a free QR reader. Readers are available for phones, tablets, laptops, desktops, and other devices. Most use the device's camera, but there are some that scan documents that are on your screen.

For resources in Science Techbook, you'll need to sign in with your student's username and password the first time you access a QR code. After that, you won't need to sign in again, unless you log out or remain inactive for too long.

We encourage you to support your student in using the print and online interactive materials in Science Techbook, on any device. Together, may you and your student enjoy a fantastic year of science!

Sincerely,

The Discovery Education Science Team

Discovery
EDUCATION

Unit 5
Senses and Survival

Bat Girl

Bats are complex organisms. What structures and behaviors do they have that help them survive? How do they communicate with one another in colonies?

Quick Code:
us4880s

© Discovery Education | www.discoveryeducation.com • Image: (a) Shahar Shabtai / Shutterstock.com, (b) annick vanderschelden photography / Moment / Getty Images

Video

Bat Girl

Discovery
EDUCATION

Think About It

Look at the photograph. **Think** about the following questions.

- How do the internal and external structures of animals help them sense and interpret their environment?

- How do senses help animals survive, grow, and reproduce?

- What role does light play in how we see?

- How do humans encode information and transmit it across the world?

Studying Bat Communication

Solve Problems Like a Scientist

Quick Code:
us4881s

Unit Project: Bat Chat

In this project, you will research bats to learn how their adaptations help them navigate and communicate.

Bat Chat

| SEP | Obtaining, Evaluating, and Communicating Information |
| CCC | Structure and Function |

Ask Questions About the Problem

You are going to create a diagram that models how bats use sound to avoid obstacles and find prey. **Write** some questions you can ask to learn more about the problem. As you learn about adaptations and senses in this unit, **write** down the answers to your questions.

Adaptation and Survival

Student Objectives

By the end of this lesson:

☐ I can model the relationships among an organism's survival, habitat, adaptations, and body systems.

☐ I can argue from evidence that plants and animals have structures and behaviors that help them survive, grow, and reproduce.

☐ I can explain how structural adaptations help organisms survive in specific environments.

☐ I can argue from evidence that multiple adaptations or organs work together in systems to help organisms survive in specific habitats.

☐ I can develop models that show how organisms respond to changes in their habitats over time.

Key Vocabulary

☐ adaptation
☐ Arctic
☐ camouflage
☐ digestive system
☐ disease
☐ ecosystem
☐ energy
☐ extinct

☐ feature
☐ hibernate
☐ migration
☐ ocean
☐ organism
☐ pollute
☐ predator
☐ prey

☐ reproduce
☐ stomach
☐ survive
☐ trait

Quick Code:
us4883s

Activity 1

Can You Explain?

How do different types of animals and plants adapt to survive cold winters?

_____

_____

_____

_____

_____

_____

_____

_____

_____

_____

Quick Code:
us4884s

Discovery
EDUCATION

Activity 2
Ask Questions Like a Scientist

Quick Code:
us4885s

Penguin Feet

Watch the video to determine why a penguin's feet are so helpful in a cold environment. Then **answer** the questions that follow.

Video

Let's Investigate Penguin Feet

SEP   **Asking Questions and Defining Problems**

Your Ideas

How do penguins' feet help them survive in cold climates?

_____

_____

_____

_____

_____

_____

_____

_____

_____

_____

Write a list of other questions you have about penguins or other animals that live in different cold environments.

_____

_____

_____

_____

_____

Activity 3
Observe Like a Scientist

Types of Adaptations

Watch the videos without taking notes.

Quick Code:
us4886s

The Arctic Fox

The Bull Shark

CCC   **Structure and Function**

Watch the videos again and **look** for specific structures and behaviors each organism has that help it survive. Then, **record** all the adaptations you observe in the table below.

Adaptations	How They Help the Animal Survive

Activity 4
Evaluate Like a Scientist

Quick Code:
us4887s

What Do You Already Know About Adaptation and Survival?

Types of Adaptations

Complete the table below by **classifying** each adaptation as structural or behavioral. **Write** S for structural adaptation and B for behavioral adaptation in the Classification column in the table.

Adaptation	Classification S = Structural B = Behavioral
Growing long, thick roots	
Flying south in winter	
Having brown fur or skin	
Taking care of offspring	
Moving in large groups	

Adaptations and Organisms

In the left column below, there is a list of adaptations found in different organisms. **Draw** a line from each adaptation to the phrase that best describes how the adaptation helps the organism survive.

Adaptation
Lightweight bones
Thick layer of fat under the skin
Camouflage
Long, straw-shaped tongue

How It Helps Organisms Survive
Helps the organism hide from enemies
Helps the organism collect food
Helps the organism fly
Helps the organism stay dry and warm

Activity 5

Analyze Like a Scientist

Quick Code:
us4888s

Adaptations

Create a bubble chart with the name of an organism in the center. Then, **work** with a partner to **add** bubbles that describe the organism's needs.

> **An Organism's Needs**

SEP Developing and Using Models

CCC Patterns

Now, **read** the text about a polar bear's adaptations.

Adaptations

Adaptations are characteristics that help living things survive and reproduce in the **ecosystem** in which they live. For example, thick, white fur is an adaptation in polar bears. It helps them stay warm in their cold, **Arctic** home. It also helps polar bears blend in with the snow as they sneak up on their prey.

In contrast, many bears that live in other habitats have darker fur. Brown bears and black bears live in forests. Their dark fur helps them stay hidden among the trees as they hunt. This type of adaptation that hides animals from a predator or their prey is called **camouflage**.

Use details from the reading to **create** an additional bubble chart for a polar bear. Each circle should include a need and an adaptation that helps the polar bear meet the need.

Polar Bear Needs

Modeling Adaptations and Survival

Now that you have thought about the polar bear and its adaptations, think about adaptations in general. **Create** a model that represents how adaptations, habitats, and an organism's survival are related. Your model can include words, images, and/or symbols.

My Model

What Are Some Examples of Adaptations in Animals and Plants?

Activity 6
Observe Like a Scientist

Quick Code: us4889s

The Panther Chameleon

Watch the video. **Look** for details about structural and behavioral adaptations. Then, **complete** the table that follows.

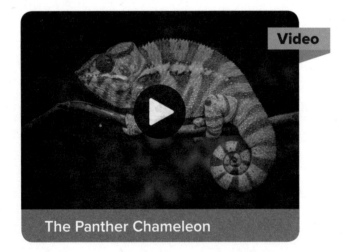

The Panther Chameleon

SEP **Constructing Explanations and Designing Solutions**

Next, **fill in** the table. First, **record** the types of adaptations you observe in the first column. Then, decide whether each adaptation is structural or behavioral. Last, determine if the adaptation helps the animal survive, grow, and/or reproduce.

Data Table: Evidence of Adaptations for Living Things

Observed Adaptation	Is the adaptation Structural (S) or Behavioral (B)?	How does the adaptation help the animal? (survive, grow, and/or reproduce)

Analyze Like a Scientist

Birds and Their Beaks

As you **read** the passage Birds and Their Beaks, **underline** any adaptations that you find. Then, **compare** what you underlined with other members of your group. When your group has agreed on the adaptations in the reading, **record** the adaptations on the data table.

Birds and Their Beaks

There are many different types of birds on Earth. All of these birds have beaks. However, their beaks can look very different. Some birds have a beak that is long and thin. Other birds have a beak that is short and wide. The shape of a bird's beak is one of the bird's traits.

All living things have traits. Traits are the features or characteristics of a living thing. Parents pass the information for traits on to their young. So, birds get the shape of their beak from their parents. Why do you think different kinds of birds have differently shaped beaks?

Birds use their beaks to help them get food. Because birds live in different places, they eat different types of food.

SEP Constructing Explanations and Designing Solutions

© Discovery Education | www.discoveryeducation.com ● Image: Discovery Communications, Inc.

|

The shape of its beak helps a bird get food in its surroundings. Woodpeckers find insects under the bark of trees. They have beaks that are sharp and strong. Woodpeckers use their beaks to peck holes in tree trunks and eat the insects underneath. A hummingbird eats nectar from inside flowers. It has a beak that is long and thin. Hummingbirds use their beaks to reach nectar deep inside flowers.

The shape of a woodpecker's beak and the shape of a hummingbird's beak help these birds survive. Birds develop different beak shapes over time. The process of developing traits that help living things survive is called adaptation. Adaptation helps living things survive in their surroundings. If living things do not develop traits through adaptation, they will die.

Observed Adaptation	Is the adaptation Structural (S) or Behavioral (B)?	How does the adaptation help the animal? (survive, grow, and/or reproduce)

DISCOVERY
EDUCATION

Activity 8
Observe Like a Scientist

Plant Adaptations

Watch the video without stopping or taking notes. **Look** for examples of plant adaptations.

Plant Adaptations

Talk Together

Now, talk together about the adaptations you saw in the video.
How do they help plants survive?

SEP **Constructing Explanations and Designing Solutions**

Activity 9

Think Like a Scientist

Plant Adaptations

In this investigation, you will explore how plants adapt to their environment.

Quick Code:
us4892s

What Will You Do?

Observe the pictures of plants that your teacher provides for you. Use the table to **describe** traits that each plant has that will help it survive in its environment. Then, **use** your observations to **answer** the questions.

Plant Environment	Characteristics of the Environment	Traits That Will Help the Plant Survive in Its Environment
Bald Cypress Swamp		
Mangrove Swamp		
Water Lily Wetland		
Conifer Forest		
Pitcher Plant Swamp		
Prickly Pear Cactus Desert		
Palm Tree Oasis		
Sugar Maple Tree Forest		
Dwarf Trees Tundra		
Acacia Trees Savannah		

Think About the Activity

What are some characteristics of plants that help them survive?

_____

_____

_____

_____

_____

Compare how plants adapt to their environment. How are they the same? How are they different?

_____

_____

_____

_____

_____

What would happen to a plant if it were placed in a different environment?

_____

_____

_____

_____

_____

SEP **Engaging in Argument from Evidence**

Activity 10

Evaluate Like a Scientist

Quick Code:
us4893s

Shaped for the Savannah

Meerkats are small mammals that live in groups on the African savannah. To best function on the grassy plains, their bodies are shaped a certain way. Meerkats also behave in ways that help them survive. **Read** the list of meerkat adaptations. **Use** the table to **classify** each adaptation as either Structural (**S**) or Behavioral (**B**). **Write** S or B in the second column of the table.

Meerkats

Meerkat Adaptation	Is the Adaptation Structural (S) or Behavioral (B)?
Strong claws	
Dig for food	
Stand guard	
Small size	
Tan and black fur	
Sleep underground	

CCC **Structure and Function**

Activity 11

Evaluate Like a Scientist

Identifying Adaptations

Quick Code:
us4894s

Explain how the adaptations help the plants in the pictures survive in their environments.

_____

_____

_____

_____

_____

_____

_____

_____

CCC **Structure and Function**

Activity 12
Analyze Like a Scientist

Body Systems

Quick Code:
us4895s

Read the text. Then, **answer** the questions that follow.

Body Systems

All organisms show individual adaptations, but how do these adaptations work together? All parts of an animal or plant work together to keep the organism alive. Without this, the animal would not survive. Different parts of an animal's body work very closely with one another to perform different jobs or functions. Parts of the body that work together are called systems. For example, animals such as humans, cows, and dogs need nutrients. To obtain and absorb these nutrients, animals need a digestive system that starts in the mouth and ends at the anus.

CCC Systems and System Models

Animal digestive systems are adapted to the types of food an animal eats. For example, the digestive system of a cow is quite different from the digestive system of a dog or a human. Cows have a digestive system that is adapted to eat grass. Grass is very difficult to digest, so cows have long digestive systems with lots of stomach-like compartments. Dogs, in contrast, eat mainly meat. Meat is much easier to process, so dogs have only one stomach and a much shorter digestive system.

Dog and Cow Digestive Systems

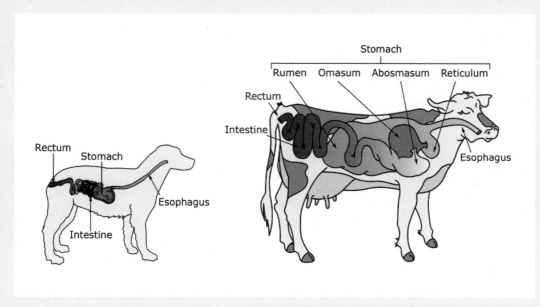

All organs and systems of organisms, whether they are animals or plants, are adapted in ways that ensure survival.

The teeth of cows and dogs are very different. Can you suggest why?

_____

_____

_____

_____

_____

_____

_____

_____

_____

_____

_____

_____

_____

_____

_____

_____

Activity 13

Observe Like a Scientist

Digestive System

Complete the interactive to learn about the digestive system. Then, **answer** the questions.

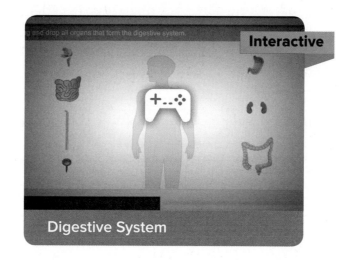

Digestive System

Why is digestion important?

_____

_____

_____

Compare what happens to food in the stomach with what happens to it in the small intestine.

_____

_____

_____

_____

Explain how the mouth helps digest food.

_____

_____

_____

_____

_____

_____

Compare and contrast the digestion that takes place in the stomach, small intestine, and large intestine.

_____

_____

_____

_____

_____

_____

Activity 14
Observe Like a Scientist

Quick Code:
us4897s

Respiratory System

Complete the interactive to learn about the respiratory system. Then, **answer** the questions.

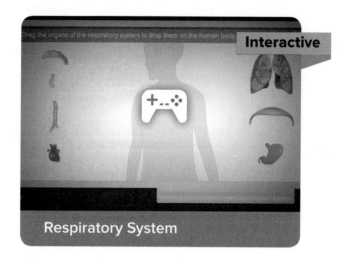

Drag the organs of the respiratory system to drop them on the human body.

Interactive

Respiratory System

Explain how the diaphragm helps us breathe in and out.

_____

_____

_____

_____

_____

CCC **Systems and System Models**

Compare the air you breathe in with the air you breathe out.

_____

_____

_____

_____

_____

How does the respiratory system get oxygen to the body cells?

_____

_____

_____

_____

_____

Why can we not hold our breath for very long?

_____

_____

_____

_____

_____

_____

Activity 15
Observe Like a Scientist

Quick Code:
us4898s

How Fish Breathe

Watch the video. **Look** for details about the structures that fish use to breathe underwater.

How Fish Breathe

 Talk Together

Now, talk together about how fish breathe. What are the similarities between the human respiration system and the fish respiration system? What are the differences?

CCC **Structure and Function**

How Do Human Activities Affect the Way Living Things Survive?

Activity 16
Observe Like a Scientist

Human Habitats

Quick Code:
us4899s

Watch the video. **Look** for details about events or situations that humans caused.

Video

Human Habitats

 Talk Together

Now, talk together about events or situations that human activity causes. What are effects of these events? What are some challenges these pose to the survival of organisms?

CCC Cause and Effect

Activity 17

Analyze Like a Scientist

Humans Change the Environment

Read the text below and **underline** evidence that humans have changed the ecosystem. Also, **circle** the impacts that human activities have on plants and animals.

Humans Change the Environment

Organisms are adapted to the ecosystems in which they live; however, that ecosystem may change. The climate may change, causing changes in temperature or in the amount of rainfall. Wildfires may change an ecosystem from a forest to dirt and eventually to grasslands. Floods may overrun areas that are usually dry. The animals in an ecosystem may also change. Predator or prey populations may increase or decrease due to these environmental changes. Or they may increase or decrease due to other factors, such as hunting or disease. Different plants may grow or die out, depending on what plants animals use for food. Humans can also change ecosystems when they farm, hunt, build, or pollute.

Sea Ice

Humans Change the Environment *cont'd*

Both plants and animals can be affected by a change in an ecosystem caused by humans. Some animals can survive by moving to another ecosystem to find what they need. Plants must rely on their seeds landing in a better place for them to survive and grow. Remember that not all changes are bad. What is bad for one organism may be good for another. Floods leave behind fertile soil for more plants to grow. Forest fires release the seeds of certain pine trees, such as the jack pine. Weed plants grow well on land that has been disturbed by human activity. Different animals and plants can move to and live in the changed ecosystem.

If an ecosystem changes, most organisms living in it will find it difficult to adapt. This is because it takes many generations for an animal or plant to adapt to its environment. Human activity rapidly changes ecosystems. Organisms cannot adapt to them fast enough. These rapid changes can cause many organisms to die. If an organism is only found in that area, it may disappear altogether and become extinct.

Activity 18
Evaluate Like a Scientist

Where Will I Survive?

Quick Code:
us4901s

Each of the moths in the images below has different adaptations. **Draw** lines to match each moth to the environment where it will best survive.

Moths

Environments

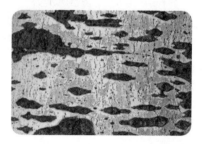

CCC Patterns

Activity 19

Record Evidence Like a Scientist

Quick Code:
us4902s

Penguin Feet

Now that you've learned about adaptation and survival, **watch** the video Penguin Feet again. You first saw this in Wonder.

Video

Let's Investigate Penguin Feet

Talk Together

How can you describe penguin feet now?

How is your explanation different from before?

SEP **Constructing Explanations and Designing Solutions**

Look at the Can You Explain? Question. You first read this at the beginning of the lesson.

Can You Explain?

How do different types of animals and plants adapt to survive cold winters?

Now, you will use your new ideas about penguin feet to answer a question.

1. **Choose** a question. You can use the Can You Explain? question or one of your own. You can also use one of the questions that you wrote at the beginning of the lesson.

My Question

2. Then, **use** the graphic organizers on the next pages to help you **answer** the question.

To plan your scientific explanation, first **write** your claim. Your claim is a one-sentence answer to the question you investigated. It answers: What can you conclude? It should not start with yes or no.

My claim:

_____

_____

_____

_____

_____

Finally, **explain** your reasoning. Reasoning ties together the claim and the evidence. Reasoning shows how or why the data count as evidence to support the claim. It may be helpful to color code your pieces of evidence and which portion of the explanation the evidence supports.

Evidence	How Evidence Supports Claim

Discovery
EDUCATION

Now, **write** your scientific explanation.

_____

_____

_____

_____

_____

_____

_____

_____

_____

_____

_____

_____

_____

_____

_____

_____

_____

STEM in Action

Activity 20

Analyze Like a Scientist

Careers and Adaptation

Read the text about reticulated glass frogs. Then, **answer** the questions that follow.

Careers and Adaptation

There is a tremendous variety of organisms living on Earth. Studying these organisms is fascinating and fun. Through research, scientists can learn how organisms adapt to their environment. Scientists can use this knowledge to help endangered species survive.

Dr. Carlos de la Rosa and other researchers work at La Selva Biological Research Station in Costa Rica. These scientists study a unique species of frog called the reticulated glass frog.

Reticulated Glass Frog

SEP **Obtaining, Evaluating, and Communicating Information**

Reticulated glass frogs have unique adaptations. They can make their bodies look like groups of their own eggs. This surprises enemies when they try to eat the eggs. The enemies find out that the eggs are actually live frogs!

Male reticulated glass frogs guard their eggs. They add water droplets to the eggs to make sure they do not dry out. The males also protect the eggs by watching for enemies. Guarding eggs is a rare adaptation in frogs.

Pollution and the cutting down of trees threaten the Costa Rican rain forests. This is a danger to the reticulated glass frogs. By learning about the frogs' adaptations, scientists can understand better how the frogs may react to these changes in their habitat.

Reticulated Glass Frogs

Reflect on what you have learned about the reticulated glass frogs. Then, **answer** the question.

Why do you think these scientists' research is important?

_____

_____

_____

_____

_____

With your group, **discuss** what you have learned about the reticulated glass frogs.

Do reticulated glass frogs have structural adaptations, behavioral adaptations, or both? Explain your answers using evidence to support your claim.

_____

_____

_____

_____

_____

_____

_____

_____

_____

_____

_____

_____

_____

_____

_____

_____

_____

Activity 21

Evaluate Like a Scientist

Review: Adaptation and Survival

Think about what you have read and seen in this lesson. **Write** down some core ideas you have learned. **Review** your notes with a partner. Your teacher may also have you take a practice test.

_____

_____

_____

_____

_____

_____

_____

 Talk Together

Think about what you saw in Get Started. **Use** your new ideas to discuss characteristics that help livings things survive and reproduce, the difference between structural and behavioral adaptations, and how human activities impact the survival of other organisms.

SEP **Obtaining, Evaluating, and Communicating Information**

Senses at Work

Student Objectives

By the end of this lesson:

☐ I can develop models that show how animals receive, process, and react to information in their environments.

☐ I can develop a model of the nervous system that includes how parts work together to carry out functions.

☐ I can explain how organs and systems work together to process and respond to input from the senses.

☐ I can plan and carry out investigations to produce evidence that the senses play a role in reaction time.

Key Vocabulary

☐ brain   ☐ receptor

☐ ear   ☐ reflex

☐ environment   ☐ senses

☐ heart   ☐ sound

☐ information   ☐ stimulus

☐ nerve   ☐ tongue

Quick Code: us4906s

Activity 1

Can You Explain?

How do animals sense and process information?

_____

_____

_____

_____

_____

_____

_____

_____

_____

_____

_____

_____

_____

Quick Code:
us4907s

Activity 2

Ask Questions Like a Scientist

Dolphin Super Senses

Watch the video. Then, **discuss** the questions and ideas you have about senses.

Quick Code:
us4908s

Video

Let's Investigate Dolphin Super Senses

SEP **Asking Questions and Defining Problems**

CCC **Systems and System Models**

Discovery
EDUCATION

Talk Together

Now, talk together about questions you have about senses. Start your questions with words such as "What," "Why," and "When."

Activity 3
Observe Like a Scientist

Super Senses

Watch the videos and **observe** how animals use their senses to help them survive. Then, **answer** the questions.

Quick Code:
us4909s

Snake Senses

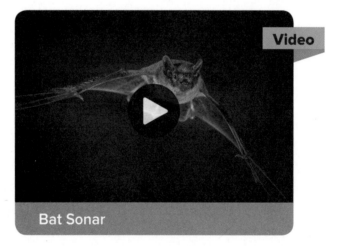

Bat Sonar

CCC **Patterns**

Elephant Feet

Snakes use heat to hunt. Why would this special sense be useful to snakes?

_____

_____

_____

_____

How do bats catch gnats in the dark?

_____

_____

_____

_____

What do elephants use to hear you from miles away?

_____

_____

_____

_____

Activity 4
Evaluate Like a Scientist

Quick Code:
us4910s

What Do You Already Know About Senses at Work?

Animal Perceptions

Sort the following senses by how the animals use them. **Write** the example in the "Hunt for Food" box if you think that is how the animals use it. **Write** the example in the "Avoid Being Food" box if you think that is how the animals use it. It is okay if you don't know all of the answers yet.

- shark's ability to smell blood
- rabbit's excellent hearing
- deer's sense of smell
- bee's ability to see ultraviolet light
- bat's echolocation

Hunt for Food	Avoid Being Food

Sensory Response

Imagine that you touch an ice cube with your index finger. Where is the information processed to tell you that it is cold? **Circle** the correct answer.

A. index finger

B. hand

C. nerve endings

D. spinal cord

E. brain

How Do Animals Sense Their Environment?

Activity 5
Observe Like a Scientist

© Discovery Education | www.discoveryeducation.com ● Image: (a) Angela Martini / Moment / Getty Images. (b) ESB Professional / Shutterstock.com, (c) Icon made by Freepik from www.flaticon.com

Using Our Five Senses

Watch the video. **Look** for information on how senses are used to make memories. Then, **discuss** with a partner.

Quick Code: us4911s

Using Our Five Senses

 ## Talk Together

Now, talk together about where you would go to create a memory. How can you use your five senses to create a strong memory of this place? Be ready to share your ideas.

CCC **Patterns**

Activity 6
Analyze Like a Scientist

Sensing the Environment

Quick Code:
us4912s

Read the text. As you read, **think** about the different senses the dog uses to find and eat his breakfast. Then, **answer** the questions that follow.

Sensing the Environment

A dog hears a can opener and smells his dog food. He then runs to his food dish. The **sound** and smell are stimuli; they cause the dog to respond. When the dog gets closer, it can see the food. As it gets to the dish, it touches the food with its mouth, and then its tongue tastes it. The dog uses its five senses to locate its breakfast. These senses are hearing, smell, sight, touch, and taste.

Dog's Breakfast

Each sense organ is different, but they all have one job: to receive information from the **environment**. Each sensory organ is specially adapted to receive this information. For example, a hand has special touch receptors that feel heat and pressure. Ears are adapted to collect sounds effectively. Eyes see by detecting light from objects.

CCC Patterns

Sensing the Environment *cont'd*

Taste receptors, located on the tongue, provide information about things to eat, such as the sweetness or saltiness of a substance. Smell provides the body with information about the odor of things using special cells in the nose. Each sense receptor is specialized to collect a particular kind of information. An **ear** cannot see, and eyes cannot taste.

Sense Organs and Senses

Organ	Name	Sense	Stimulus
	Eye	Sight	Detected Light
	Nose	Smell	Chemical Substances
	Skin	Touch	Touch, Pressure, Pain, Cold and Heat
	Ear	Hearing and Balance	Sound
	Tongue	Taste	Chemicals

Each sense organ collects information on a single type of stimulus. What pattern do you observe in the Sense Organs and Senses table?

_____

_____

_____

_____

Activity 7

Observe Like a Scientist

What Is the Nervous System?

Quick Code:
us4913s

Watch the video. **Look** for the parts of the nervous system and how these parts interact.

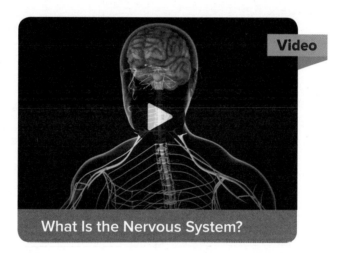

What Is the Nervous System?

 Talk Together

Now, talk together about how the nervous system is like a model of a cable communication network.

SEP Developing and Using Models

Use the chart to **compare** the different parts and to **identify** similarities and differences.

Nervous System	Cable Communication Network

Analyze Like a Scientist

Pizza and the Nervous System

Read the text. Then, **complete** the activity that follows.

Quick Code:
us4914s

Pizza and the Nervous System

In mammals, such as elephants, humans, and dogs, the nervous system is made up of the **brain**, the nerves, and the sense organs. The brain is connected to a big **nerve** that runs through the backbone, called the spinal cord. Smaller nerves branch off this main nerve and go to different parts of the body. These nerves branch further and go to individual sensors, muscles, and other body cells. A few nerves, such as those from the eyes and **heart**, connect directly to the brain.

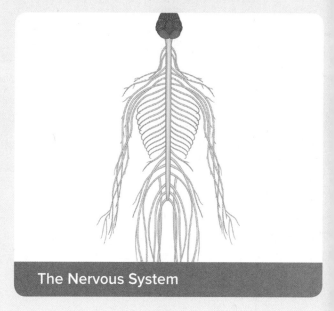

The Nervous System

CCC Systems and System Models

Pizza and the Nervous System *cont'd*

The sense organs receive information from the environment. Nerves in the body connect the sense organs to the brain. Nerves are constantly receiving information from the senses and sending the information to the brain. For instance, if you smell a pizza, that information is detected by your nose. Then, nerves at the back of the nose send a specific signal to your brain. The signals travel as electrical impulses from the sense organ along the nerves to the brain. Once the information about the smell reaches the brain, the brain can determine what to do with that information, including how to react.

Identify and **list** the parts of the nervous system.

_____

_____

_____

_____

_____

_____

_____

Activity 9
Observe Like a Scientist

Processing Sensory Information

Quick Code: us4915s

Watch the video. **Look** for what organs do in response to stimuli.

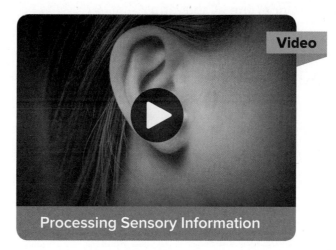

Processing Sensory Information

 Talk Together

Now, talk together about how organs and systems work together to process and respond to different stimuli.

CCC **Systems and System Models**

Activity 10

Evaluate Like a Scientist

Quick Code:
us4916s

Sensing the Environment

The following paragraph describes a person making observations and acting on them. Which sentences state **incorrect** information about how this process happens? **Underline** all that apply.

Kwan is walking to school. As he approaches the corner, he hears a sound. The sound is processed in his ear. The nerves send information about the sound to his brain. Using past experiences, the brain determines that the sound is a moving car. The brain sends a signal to Kwan's body to stop moving. The muscles in Kwan's legs make him stop on the corner. Kwan is thankful that his nervous system's muscles, nerves, and brain work together to keep him safe.

© Discovery Education | www.discoveryeducation.com • Image: Angela Martini / Moment / Getty Images

SEP   **Engaging in Argument from Evidence**

How Can Different Parts of the Body Work Together as a System?

Activity 11

Observe Like a Scientist

Quick Code:
us4917s

Nerves

Test your reaction time. As you perform the tests, **record** your data in the table. Then, **answer** the questions that follow.

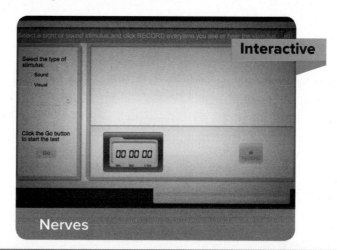

Nerves

Data:

Trial	Time	Trial	Time

SEP **Planning and Carrying Out Investigations**

How do nerves function when you see something?

_____

_____

_____

_____

_____

What things might affect our reaction time to the sights and sounds presented in this interactive?

_____

_____

_____

_____

_____

_____

Activity 12

Investigate Like a Scientist

Quick Code:
us4918s

Hands-On Investigation: Reaction Time

In this investigation, you will **examine** reaction time for catching a meter stick that is dropped. In the first part of the investigation, you will use the sense of sight to see when the meter stick is dropped. In the second part, you will use the sense of sound, listening for a signal to know the meter stick was dropped.

Make a Prediction

Which sense will have the faster reaction time, sight or sound? Explain your prediction.

_____

_____

_____

_____

_____

_____

_____

_____

_____

SEP Planning and Carrying Out Investigations

CCC Systems and System Models

What Will You Do?

1. Measure reaction time catching a meter stick using the sense of sight three times.

2. Measure reaction time catching a meter stick using the sense of sound three times.

3. Record your observations on the Reaction Time Data Table below.

4. Calculate the average distance for each.

5. Use the Meters/Second Conversion Chart to convert your average distance to reaction time. Record the time in the Reaction Time Data Table.

Meters/Second Conversion Chart

Distance (cm)	Distance (in)	Time
5	2	.10 sec
10	4	.14 sec
15	6	.17 sec
20	8	.20 sec
25.5	10	.23 sec
28	11	.25 sec
43	17	.30 sec
61	24	.35 sec
79	31	.40 sec
99	39	.45 sec
122	48	.50 sec
176	69	.60 sec

What materials do you need? (per group)

- Meter stick
- Chair
- Calculator, solar

Reaction Time Data Table

Student	Trial 1	Trial 2	Trial 3	Average Distance	Reaction Time

Think About the Activity

Which resulted in the quicker reaction time: auditory or visual stimuli? How was the information received and processed in each of these examples?

_____

_____

_____

_____

What senses were used in this investigation? How was the information processed?

_____

_____

_____

_____

_____

_____

_____

Was there a difference in reaction time between seeing the ruler drop and being told it was dropped? Use what you have learned to explain your answer.

_____

_____

_____

_____

What are two examples of when reaction time is important?

_____

_____

_____

_____

Why was it important to do multiple trials for each person?

_____

_____

_____

_____

Activity 13
Observe Like a Scientist

Quick Code:
us4919s

How the Nervous System Works

Watch the video. **Look** for the difference between sensory and motor input and how messages travel to and from the brain.

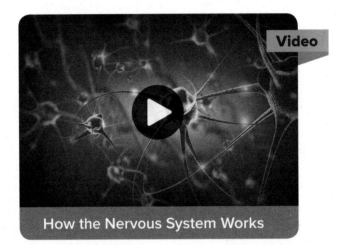

How the Nervous System Works

Talk Together

Now, talk together about how the parts of the nervous system are connected. Be ready to share your ideas.

SEP **Engaging in Argument from Evidence**

Activity 14
Evaluate Like a Scientist

Quick Code:
us4920s

The Nervous System, Job of the Nervous System, and Describe the Nervous System

The Nervous System

Look at the following images. Which of these are part of the nervous system? **Circle** all that apply.

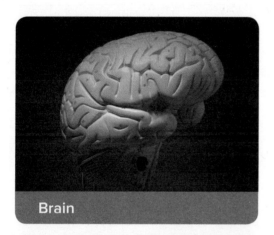

Brain

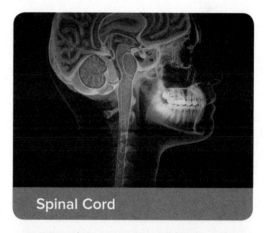

Spinal Cord

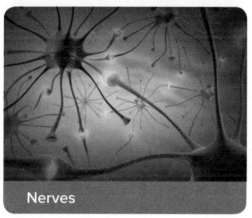

Nerves

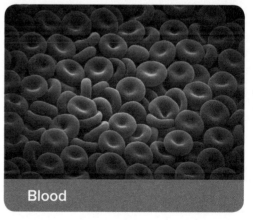

Blood

CCC Systems and System Models

Job of the Nervous System

Think about what you have learned about the nervous system and **explain** what the parts of the nervous system can do together that the individual parts cannot do alone. **Share** your sentence with a partner.

_____

_____

_____

_____

_____

_____

_____

_____

_____

_____

_____

_____

_____

_____

_____

_____

Describe the Nervous System

Read the sentences that describe the nervous system. **Write** the correct term from the word bank in each blank. You will not use all the terms.

heart	brain	nerves
blood	central nervous system	endocrine system
voluntary	involuntary	peripheral nervous system

1. The _____ is like the command center for your body.

2. _____ send(s) messages to the brain.

3. The brain is part of the _____.

4. Breathing is a(n) _____ action.

5. Nerves in the fingers are part of the _____.

Activity 15
Analyze Like a Scientist

Quick Code: us4921s

Your Nervous System

Choose two or three of the following reading sections: Your Sensational Senses, Your Nervous System at Work and Play, or Messages in Motion. As you read your chosen passages, **look** for clues to how the nervous system functions. Then, **complete** the activity that follows.

Your Sensational Senses

Take a minute to think about what it's like to be YOU. Look around—what do you see? What sounds do you hear? What do you smell, taste, or feel?

Now try to answer this question in just one word: What five things help you experience the world? (Read on to find the answer!)

Take a Look

About 80 percent of what you know about your surroundings comes from your *eyes*. When light enters your eyes, signals are sent along nerves to your brain. Then, your brain tells you what you see.

SEP Obtaining, Evaluating, and Communicating Information
CCC Systems and System Models

Your eyes help you find food, recognize people, and stay safe. If your brain thinks something you see is dangerous, it may tell your muscles to move you away quickly!

Listen Up

Your ears help you hear sounds. When your ears pick up vibrations in the air, they send messages along nerves to your brain. Your brain recognizes the signals as sounds.

Sometimes when your brain gets a message, you choose to do something. You might dance when you hear music or giggle when you see something funny. But often, your brain sends messages that you don't control.

What's That Smell?

Has your stomach ever started rumbling when you smelled food? When your nose catches scents from the air, it lets your brain know. If your brain recognizes the smell as food, it might tell your stomach to get ready for action!

The smell of food cooking might make you hungry.

Rotten food, sewage, or certain chemicals can make you sick. When you smell one of these things, your brain might send messages that make you gag or move away.

Mmm . . . Tasty

When something touches your tongue, signals are sent to your brain that help it decide whether the food is good to eat. Spoiled food usually tastes bad, so your brain might tell you to spit it out. Your nose and tongue work together to keep dangerous materials out of your body.

Feeling Fine

Did you know that you have several pounds of skin on your body? Many nerves end at your skin, making it very sensitive. Your skin lets you feel things like heat, cold, pressure, and pain.

Certain parts inside you, such as your stomach and bladder, have some feeling, too. Your brain might determine that you are full or need to use the bathroom based on pressure in these organs.

Now do you know the answer to the question? SENSES! Your five senses (sight, hearing, smell, taste, and touch) work with your brain and nerves to help you experience the world!

Your Nervous System at Work and Play

Imagine that you are playing baseball. Your friend pitches the ball and you swing the bat—WHACK! You sprint around the bases. It's a home run!

After scoring, you take a break. Sweat runs down your cheek. Just then, you hear your friend's mom call from the house, "Lemonade!"

Your Nervous System

Did you know that your nervous system helped you play baseball? Your nervous system includes your brain, spinal cord, and nerves. Your spinal cord and nerves are paths that carry messages between your brain and different body parts.

When you hear, smell, feel, see, and taste, you get messages about the world. These messages travel along nerves, up your spinal cord, and to your brain. Your brain uses this information to decide what your body should do next.

Batter up! When you were at bat, your eyes saw the ball coming. Your hands felt the ball hit the bat. Your ears heard the WHACK! Then, you saw the ball fly over the fence. All of these messages were sent to your brain.

NERVOUS SYSTEM

brain

spinal cord

nerves

Your brain, spinal cord, and nerves send messages around your body.

Your brain used this information. When your brain learned that the ball was coming, it sent messages down your spinal cord. The messages moved along nerves to your arm muscles. Then, your arm muscles helped you swing the bat to hit the ball.

Your nervous system was doing many other jobs, too. Remember feeling sweat on your cheek? Your muscles made heat when you ran, so your brain sent messages to your skin to sweat. Sweating helps cool your body.

And don't forget another great way to cool off—drinking an ice-cold glass of lemonade!

Messages in Motion

The signals your brain receives from your senses help you recognize the things around you, such as this sea star.

Imagine being at the beach. Listen to waves crash and feel smooth sand between your toes. Watch the water splash against rocks and taste the salty spray on your lips. Inhale the sweet aromas of the ocean breeze.

Can you name any senses that would make this experience possible? If you mentioned sight, smell, taste, touch, and hearing, you're correct! But senses don't work alone. They rely on your brain and other parts of your nervous system to help you experience the world.

Senses: Gathering Information

Your sense organs include your eyes, ears, tongue, nose, and skin. When they collect information, signals are sent along nerves to your brain. Then, your brain can determine what to do next and send new messages throughout your body.

Your eyes send signals to your brain that help it form images of your surroundings. Your ears let you know what sounds are present. Your nose helps you smell and works with your tongue to help you taste. Your skin and other structures help you feel things like heat, cold, pressure, and pain.

Brain: Responding to Information

Your brain responds in different ways to information from your senses. Sometimes, you can control the responses. Other times, your brain sends messages directly to body structures without you thinking about it.

Think again of being at the beach. If your eyes notice something moving in the water, signals would travel to your brain. If you realize it is a sea star, you might decide to pick it up. So, your brain would send messages along nerves to muscles. These signals would cause the right muscles to contract or relax for you to pick up the sea star.

Suppose clouds move in and the ocean breeze becomes cool. When your skin feels the temperature drop, your brain might send messages back to your skin. You might get goose bumps, causing hairs to stand up and trap heat around your skin. You didn't choose for this to happen—it was an autonomic reaction.

Your brain, spinal cord, and nerves help send and receive messages around your body. Many animals have the same five senses as humans. Your brain controls your internal organs, such as your heart and lungs, in the same way. It is constantly sending and receiving messages from inside your body to keep your organs functioning properly. And you never have to think about it!

Reflexes: Bypassing the Brain

Sometimes, your body needs to react quickly to a sensation. Remember the sea star in the water? Suppose it had been a prickly sea urchin instead. If you touched it, you would have felt a sharp sting. Did you know your fingers might have even jerked back slightly before it hurt? This is called a reflex.

When your finger gets pricked, messages move along nerves, up your spinal cord, and to your brain to let you know your body is in pain. However, other signals from your finger follow a different, shorter route. When the signals reach the spinal cord, messages are immediately sent back to your hand to pull away. Reflexes help you move quickly from danger.

Your brain and nervous system communicate with all parts of your body. They help you move, stay healthy, and experience the world around you.

Which passages did you read?

_____

_____

Using information from the passages, **explain** how the nervous system works.

_____

_____

_____

_____

_____

_____

_____

_____

_____

_____

_____

_____

_____

_____

Activity 16

Record Evidence Like a Scientist

Quick Code:
us4922s

Dolphin Super Senses

Now that you have learned about how animals use their senses to understand and react to their environment, watch the video about dolphin senses again. You first saw this in Wonder.

Video

Let's Investigate Dolphin Super Senses

Talk Together

How can you describe dolphin super senses now?
How is your explanation different from before?

SEP **Constructing Explanations and Designing Solutions**

© Discovery Education | www.discoveryeducation.com • Image: (a) Angela Martini / Moment / Getty Images, (b) Andrea Izzotti / Shutterstock.com. (c) Icon made by Freepik from www.flaticon.com

Look at the Can You Explain? question. You first read this question at the beginning of the lesson.

Can You Explain?

How do animals sense and process information?

Now, you will use your new ideas about senses to answer a question.

1. **Choose** a question. You can use the Can You Explain? question or one of your own. You can also use one of the questions that you wrote at the beginning of the lesson.

My Question

2. Then, **use** the graphic organizers on the next pages to help you **answer** the question.

To plan your scientific explanation, first **write** your claim. Your claim is a one-sentence answer to the question you investigated. It answers: What can you conclude? It should not start with yes or no.

My claim:

_____

_____

_____

_____

_____

_____

Record the reasons and evidence to support your claim in the graphic organizer.

Reasoning	Evidence

DISCOVERY
EDUCATION

Now, **write** your scientific explanation.

_____

_____

_____

_____

_____

_____

_____

_____

_____

_____

_____

_____

_____

_____

_____

_____

STEM in Action

Quick Code:
us4923s

Activity 17

Analyze Like a Scientist

Careers: Become a Neuroscientist

With your partner, take turns **reading** one paragraph at a time, out loud. When you are the listener, **share** with your partner one thing you found interesting from the paragraph.

Careers: Become a Neuroscientist

Neuroscientists study how the brain and nerves work. Their job is to find out how the 100 billion nerve cells in the brain grow and communicate with one another and with the nerves of the body. The brain and the nerves of the body work together so the body does what it needs to do to stay alive. Neuroscientists try to answer questions like: How does the brain help us read, communicate, and feel? How do we remember people in our

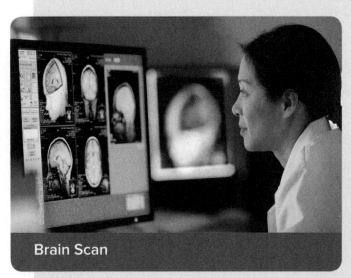

Brain Scan

SEP Obtaining, Evaluating, and Communicating Information

CCC Cause and Effect

CCC Systems and System Models

lives or the things that happened a long time ago and things that happened only yesterday? How do we learn a new language?

ALS and Basketball

Neuroscientists help people whose brain and nerves do not work well. Amyotrophic lateral sclerosis, or ALS, affects the voluntary actions of the body. Voluntary actions are movements of your body that you need to think about to make them happen. For example, if you put cheese in your mouth, your brain tells you that you have food in your mouth. It also tells you that you need to chew it. Your mouth will not chew your food without your brain telling your mouth to do its job. When a person has ALS, the nerves that make the body move do not work well or have died. Scientists are now doing experiments to better understand how the nerves die. They are also testing substances and other treatments that can be used to help people with this disease.

Another disease that affects how the brain works is Alzheimer's disease. This disease destroys the part of the brain that helps people remember things. This is not the same as being forgetful. People who have this disease might even forget people who are important to them. Because of this, they may become confused all the time. The nerve cells of a person who has this disease cannot communicate with one another and the brain. Neuroscientists have not found a cure for this disease. They are working hard on studying the genes that cause the disease. If they know what genes cause the disease, they can work with other scientists to create medicine that can help cure or prevent the disease.

Genes

How does one become a neuroscientist? A student would need to study hard when he or she is in elementary, middle school, and high school. Then, in college, a student can choose what type of neuroscientist he or she wants to be. Some neuroscientists study how the brain develops and grows. Others study how the brain affects how people and animals behave and feel. Still others study diseases caused by nerve damage, so they can help sick people.

Mahmood Amiry-Moghaddam came to Norway as a refugee from Pakistan when he was a teenager. When he arrived in Norway, he studied biology and medicine in college. Then, he got a PhD in neuroscience. He studies how brain and nerve cells work. This research can help find cures or treatments for diseases that are caused when brain and nerve cells are not working well.

Now that you know what neuroscientists do and how they help people, would you like to become one?

Test Your Memory

Think about what you have learned about senses and the nervous system. Then, **answer** the question.

How are you using your nervous system to remember people you know or things that have happened in the past year? How is this process different when a person has Alzheimer's disease?

_____

_____

_____

_____

_____

_____

_____

_____

_____

_____

_____

_____

_____

Activity 18

Evaluate Like a Scientist

Quick Code:
us4924s

Review: Senses at Work

Think about what you have read and seen in this lesson. **Write** down some core ideas you have learned. **Review** your notes with a partner. Your teacher may also have you take a practice test.

_____

_____

_____

_____

_____

_____

_____

_____

_____

_____

_____

_____

SEP Obtaining, Evaluating, and Communicating Information

Talk Together

Think about what you saw in Get Started. Use your new ideas to discuss how we get information from our senses and how using their senses helps organisms survive.

Light and Sight

Student Objectives

By the end of this lesson:

☐ I can describe how light transfers energy across distances.

☐ I can develop a model that describes how the behavior of light enables the eye to see objects.

Key Vocabulary

☐ heat

☐ light

☐ matter

☐ opaque

☐ organ

☐ pupil

☐ reflect

☐ refract

☐ seismic

☐ skin

☐ thermal energy

☐ transparent

Quick Code:
us4926s

Activity 1

Can You Explain?

What needs to happen for you to see an object in low-light areas?

_____

_____

_____

_____

_____

_____

_____

_____

_____

_____

Quick Code:
us4927s

Discovery EDUCATION

Activity 2

Ask Questions Like a Scientist

Quick Code:
us4928s

Hunting with Night Vision

Watch the video Little Red Bats. Then, **complete** the activity that follows.

Little Red Bats

Circle the statement that is false.

Bats hunt at night.

All bats are blind.

A bat can use echolocation to find its way in the dark.

All healthy bats can fly.

SEP **Asking Questions and Defining Problems**

Watch the video. Then, **discuss** what you notice about how your own vision works at night.

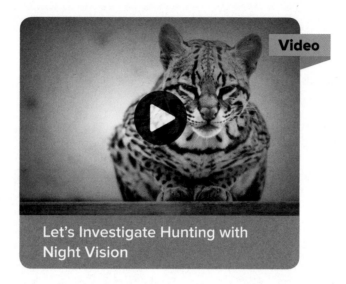

Let's Investigate Hunting with Night Vision

Talk Together

Now, talk together about how your vision works in the dark. As you talk, consider your ideas about what needs to happen to help you see in the dark.

Activity 3
Observe Like a Scientist

Quick Code:
us4929s

Hunting in the Dark

Watch the videos, **examine** the images, and **look** for how and why humans and animals adapt to dark places.

Tarsiers Hunt at Night

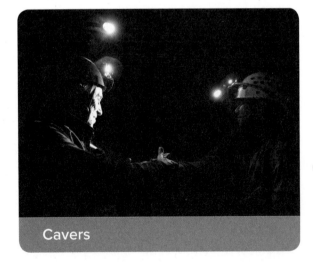

Cavers

Cat Eyes in the Dark

CCC **Energy and Matter**

After watching the videos and studying the images, **complete** the chart to **identify** the similarities among the abilities of humans, cats, and tarsiers to adapt or see in dark places.

Adapting to the Dark		
Humans	Cats	Tarsiers

Activity 4
Evaluate Like a Scientist

Quick Code:
us4930s

What Do You Already Know About Light and Sight?

Sources of Light

Look at the pictures below. **Circle** the pictures that show sources of light. Be careful! Remember that a source of light is not something that reflects light. It is something that gives off its own light.

© Discovery Education | www.discoveryeducation.com ● Image: (a) Witsanu Thangsombat / EyeEm / Getty Images, (b–c) Paul Fuqua, (d) pinkypills/Fuse / Getty Images, (e) Tetra Images / Getty Images, (f) Jorg Greuel / DigitalVision / Getty Images, (g) Jennifer Smith / Moment Open / Getty Images, (h) Gregory Alonso / EyeEm / Getty Images, (i) Pixabay

How We See

Look at the images below. **Circle** the one that best shows what happens when you see a red ball.

A.

B.

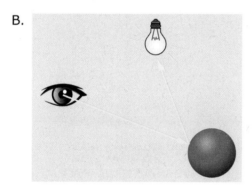

C.

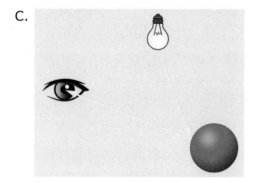

Activity 5
Analyze Like a Scientist

Quick Code:
us4931s

Light Is Energy

Read the text. Then, with a partner, **discuss** how you can illustrate the information in each paragraph. Once you agree, **sketch** the illustration you chose for each paragraph.

Light Is Energy

What do you notice when you move from a shady area into the sunlight? How does the sunlight feel on your **skin**? You may feel your skin warming. If you spend a lot of time in the sunlight, you also may notice your skin feeling hot, or even burning. Have you ever hung clothes in the sunlight to dry? You may have noticed that objects dry faster in sunlight than in shade. Sunlight can cause all of these changes because **light** is a form of energy.

But what is light? To learn more about it and how it behaves, you must know what it is. Light is a form of energy that can be seen. As you learn more about light, keep in mind that it is a visible form of energy that travels in the form of a wave.

CCC **Energy and Matter**

We know that light is a form of energy because of the changes it can cause. For example, solar panels can transform light energy from the sun into electricity to power our homes. Energy can also transform from one form to other forms. We observe that light energy transforms to **thermal energy** when objects warm up in sunlight. We feel thermal energy as **heat**.

Paragraph 1	
Paragraph 2	
Paragraph 3	

Activity 6
Observe Like a Scientist

Quick Code:
us4932s

Lighting a Campfire

Look at the image and **answer** the questions. Then, **discuss** your answers with a partner.

Lighting a Campfire

CCC **Energy and Matter**

What energy change is taking place here?

_____

_____

_____

_____

_____

_____

What other forms of energy can be transformed into light energy?

_____

_____

_____

_____

_____

_____

Activity 7

Investigate Like a Scientist

Hands-On Investigation: Light Observations

In this investigation, you will use different objects to observe how light transfers energy.

Make a Prediction

Which solar-powered object do you think will provide the most evidence that light can transfer energy? Explain your prediction.

_____

_____

_____

SEP   Engaging in Argument from Evidence

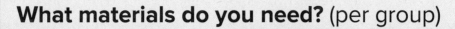

What materials do you need? (per group)

- Calculator, solar

- Radiometer

- Flashlight

- Shoebox with two holes in the lid

- Object that can fit in the box

- Thermometers, plastic

- Clear plastic containers, 2

- Plastic cups, 9 oz

- Ice cubes

- Clamp on light with shade

- Batteries, size D

- Frosted light bulb

- Camera or video recorder to record demonstrations (optional)

What Will You Do?

1. Use the materials at each station to plan and conduct a demonstration to show how light transfers energy.

2. Collect evidence about each object when it is in the light and when it is not.

As you collect your evidence, **record** your observations in the chart below.

Station	Materials	Directions for using the materials (include a sketch if it would be helpful)	How this demonstrates that light transfers energy
1			
2			
3			
4			
5			
6			

Think About the Activity

If you could use only one demonstration to convince someone that light transfers energy from place to place, which would you use? Why?

_____

_____

_____

_____

Choose one of the demonstrations. If you could have other materials or more time, how could you improve it?

_____

_____

_____

_____

Analyze Like a Scientist

Quick Code: us4934s

Light Transfers Energy

Read the text. **Consider** ways light transfers energy. After reading, you will work with your group to **choose** a system. Then, **draw** a model of your chosen system and **describe** how energy enters it and leaves it.

Light Transfers Energy

We can observe how light transfers energy. You are observing light transferring energy as you look at the computer screen. The screen

converts electricity into light that reaches your eyes. A lamp on the ceiling produces light that travels across the room. Do you have a solar-powered calculator or toy? What happens when you cover the solar cell on it and block the light from reaching it? Or observe how light transfers from a flashlight to the solar cells. The solar cells then transform light energy to electrical energy that powers the device.

SEP Engaging in Argument from Evidence

CCC Energy and Matter

My system:

Activity 9
Evaluate Like a Scientist

Quick Code:
us4935s

Evidence for Light Energy

Explain how light is a form of energy. Use three observations from your everyday life to support this.

_____

_____

_____

_____

SEP   Engaging in Argument from Evidence

Activity 10

Investigate Like a Scientist

Quick Code:
us4936s

Hands-On Investigation: Reflection

In this activity, you will work with a group to demonstrate how light can be reflected. Then, you will compare how well different materials reflect light.

Make a Prediction

Which objects do you think will reflect light best? **Write** and **explain** your hypothesis.

_____

_____

_____

What Will You Do?

1. Demonstrate the reflective properties of three objects.

2. Compare the reflective properties of three objects.

SEP **Planning and Carrying Out Investigations**

CCC **Cause and Effect**

What materials do you need? (per group)

- Flashlight
- Batteries, size D
- Fake fur
- Plastic block
- Wooden blocks
- Small mirror
- Piece of shiny metal
- Piece of marble
- Paper
- Painted metal
- Other materials

Material	Observations	Is this what you expected to happen?

Think About the Activity

Review your hypothesis. Did the results of the investigation provide evidence that supported your hypothesis? Or did it provide evidence against your hypothesis? Describe how you know.

_____

_____

_____

_____

_____

_____

Based on your results, which types of materials reflect light the best? Which reflect light poorly? Explain your answer.

_____

_____

_____

_____

_____

_____

How could you provide more evidence for your answer to the previous question? **Describe** another investigation that you could conduct.

_____

_____

_____

_____

_____

_____

Draw a picture of your results showing the paths of light rays and using angles to describe the reflections.

How does the path that light reflects relate to the movement of energy?
Is light itself a form of energy?

_____

_____

_____

_____

_____

_____

Draw a picture that shows how light travels from the sun and then bounces off a mirror.

What additional questions do you have about the way materials reflect light? Choose one of your questions and explain how you could learn information to help you answer the question.

_____

_____

_____

_____

_____

_____

_____

_____

Discovery EDUCATION

Analyze Like a Scientist

Light Strikes Matter

Read the text and **look** at the images. Then, **answer** the question that follows.

Light Strikes Matter

Light energy is released from a light source. For example, the sun, a light source, emits light energy. This travels as waves, in straight lines, in all directions away from the sun. A tiny portion of this energy reaches Earth. When it reaches Earth, it strikes **matter**. When light waves hit an object, some of the energy is absorbed. Some of the energy may go through the object. Some of the energy bounces, or reflects, off the object's surface. You can examine these behaviors of light by observing different objects. Some objects, including your body, make shadows. This happens because all of the light that hits your body either bounces off or is absorbed. None of the light passes through you. Objects that light cannot pass through are called **opaque**. **Transparent** objects or substances, such as air, water, windows, and lenses, allow light to pass through, which is why you can see through them.

CCC **Energy and Matter**

Refraction

When light hits transparent matter, it slows down and changes direction. This process is called refraction. If you have ever been swimming in a clear pool, you will have observed refraction. Refraction is why, once in the water, your feet never appear to be where they should. Lenses work by bending light by refraction. Lenses like those in glasses, binoculars, and cameras are designed to bend or **refract** light in specific ways.

When light hits an opaque object, some of it is absorbed. When this happens, light energy is converted to heat (thermal energy). This explains why objects get hot when placed in intense sunlight. The rest of the energy bounces, or reflects, off. How the light is reflected depends upon the smoothness of the surface. If the surface is a polished mirror, the rays **reflect** off differently than from a painted surface, which is slightly rough.

Discovery
EDUCATION

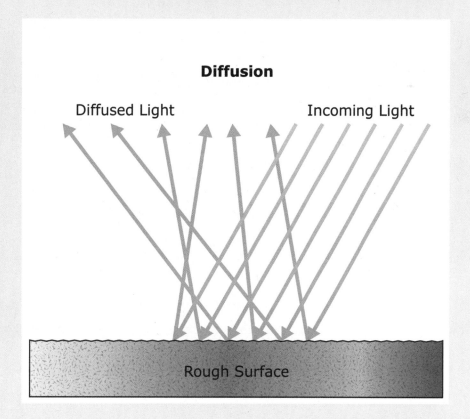

Reflection

Reflected Light   Incoming Light

Smooth Surface

Diffusion

Diffused Light   Incoming Light

Rough Surface

Your older sister dropped her cell phone and now the screen has a few cracks. How do you predict that light will reflect off of the screen compared to before it was broken?

_____

_____

_____

_____

_____

Activity 12

Observe Like a Scientist

Refraction: The Bending of Light Rays

Quick Code:
us4938s

Watch the video. **Look** for how light is associated with a bent pencil.

Refraction: The Bending of Light Rays

Talk Together

Now, talk together about what can happen to light when it hits a surface and the difference between reflection and refraction.

CCC **Energy and Matter**

How Do We See Objects?

Activity 13

Think Like a Scientist

Optical Illusions

Quick Code:
us4939s

In this investigation, you will work in groups to find an optical illusion and develop a question to ask. Then, you will survey class members to see what they observe in the illusion.

What materials do you need? (per group)

- Optical illusions

SEP Planning and Carrying Out Investigations
CCC Patterns

What Will You Do?

1. Choose an optical illusion.

2. Develop a question to ask.

3. Survey the class to see what they observed.

4. Record findings in the chart below.

Survey Data

Question: _____

_____

Yes	No

Think About the Activity

What patterns did you notice in the optical illusions?

_____

_____

_____

_____

_____

Was everyone able to see the same thing? What do you think caused the differences?

_____

_____

_____

_____

_____

What role did the eye and the brain play in the optical illusion?

_____

_____

_____

_____

_____

© Discovery Education | www.discoveryeducation.com ● Image: Witsanu Thangsombat / EyeEm / Getty Images

Activity 14
Evaluate Like a Scientist

Sight Model

Imagine using a bouncing ball to model how we see reflected light. **Choose** a common object to represent eyes in the model. **Explain** how you could use the model to demonstrate how we see reflected light from objects. **Include** all of the following in your response:

- Summarize which parts of your model represent how we see light from reflected objects.

- Relate your model to the way in which we see light reflected off objects.

- Explain what you learned about reflection and sight from your model.

_____

_____

_____

_____

_____

_____

_____

_____

SEP **Developing and Using Models**

Activity 15

Record Evidence Like a Scientist

Hunting with Night Vision

Now that you have learned about how vision works, **watch** the video Let's Investigate Hunting with Night Vision again. You first saw this in Wonder.

Quick Code:
us4941s

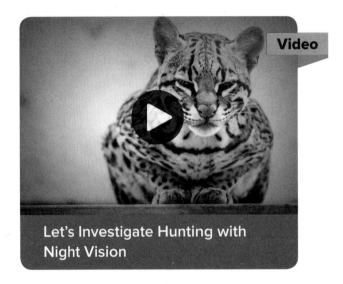

Video

Let's Investigate Hunting with Night Vision

Talk Together

How can you describe hunting with night vision now?
How is your explanation different from before?

© Discovery Education | www.discoveryeducation.com • Image: (a) Witsanu Thangsombat / EyeEm / Getty Images, (b) PicksArt / Shutterstock.com, (c) Icon made by Freepik from www.flaticon.com

SEP   Constructing Explanations and Designing Solutions

Look at the Can You Explain? question. You first read this question at the beginning of the lesson.

 ## Can You Explain?

What needs to happen for you to see an object in low-light areas?

Now, you will use your new ideas about how light and vision work to answer a question.

1. **Choose** a question. You can use the Can You Explain? question or one of your own. You can also use one of the questions that you wrote at the beginning of the lesson.

My Question

2. Then, **use** the graphic organizer on the next pages to help you **answer** the question.

To plan your scientific explanation, first **write** your claim.

My claim:

_____

_____

_____

_____

Record the reasons and evidence to support your claim in the graphic organizer.

Reasoning	Evidence

© Discovery Education | www.discoveryeducation.com   ●   Image: Witsanu Thangsombat / EyeEm / Getty Images

Discovery EDUCATION

Now, **write** your scientific explanation.

_____

_____

_____

_____

_____

_____

_____

_____

_____

_____

_____

_____

_____

_____

S T E M in Action

Quick Code:
us4942s

Activity 16

Analyze Like a Scientist

How Do Optometrists Help Us See?

Read the text. Then, **complete** the Eye Imperfections activity. After the activity, **discuss** your answer with your partner.

How Do Optometrists Help Us See?

When you see, what path does the light take to get to your eye? What happens when the light reaches your eye? Did you know that there is a lens just inside your eye that focuses the light onto the retina in the back of your eye?

When the lens focuses light, it redirects the light so that it all goes to a point. Think about a magnifying glass. It can take the sun's rays and concentrate them on a single point. Or it can take the light that is bouncing off something small, like an insect, and focus it on your eye.

SEP Obtaining, Evaluating, and Communicating Information

If the lens in your eye does not focus properly, you may have blurry vision. Some people are not able to see objects that are far away. Others have a hard time seeing objects that are near.

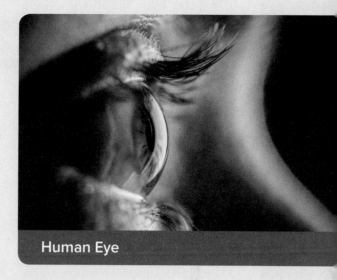

Human Eye

Optometrists like Dr. Patricia Perez can test your eyes to determine whether the lens is focusing properly. By using a series of tests, she can determine how to correct your vision: perhaps with glasses or contact lenses, or maybe even using laser surgery. To learn how to become an optometrist, after high school, Dr. Perez studied for four years of college and then four years at an optometry school. Specifically, she received her doctorate from the Illinois College of Optometry. In so doing, she learned how to prevent blindness, treat eye diseases, and correct vision. Dr. Perez currently runs her own practice and lives in Chicago. If you want to become an optometrist like Dr. Perez, you need to study hard and learn a lot about science.

Eye Imperfections

Some people have difficulty seeing objects near them, while other people have difficulty seeing objects far from them. Some have difficulties distinguishing between colors.

Others cannot see in their peripheral vision. Given what you know about eyes, **create** a test to look for one of these difficulties.

_____

_____

_____

_____

_____

_____

_____

_____

_____

_____

_____

Activity 17

Evaluate Like a Scientist

Review: Light and Sight

Think about what you have read and seen in this lesson. **Write** down some core ideas you have learned. **Review** your notes with a partner. Your teacher may also have you take a practice test.

_____

_____

_____

_____

_____

_____

_____

 Talk Together

Think about what you saw in Get Started. **Use** your new ideas to discuss how animals use light to see, and how this helps them survive.

SEP **Obtaining, Evaluating, and Communicating Information**

Communication and Information Transfer

Student Objectives

By the end of this lesson:

☐ I can compare solutions that use patterns to transfer information.

☐ I can develop models to compare analog and digital signals.

☐ I can develop a model of a communication system with many parts that work together to transfer information from one place to another.

☐ I can argue, using evidence, that light and sound transfer energy across distances.

☐ I can design, test, and evaluate models of information-transfer systems that can send and receive coded information.

Key Vocabulary

☐ air

☐ analog

☐ antenna

☐ code

☐ digital

☐ electromagnetic spectrum

☐ system

Quick Code: us4945s

Activity 1

Can You Explain?

How do humans use light and other electromagnetic radiation to send and receive information?

_____

_____

_____

_____

_____

_____

_____

_____

_____

_____

Quick Code:
us4946s

| **⊕DISCOVERY** EDUCATION

Activity 2
Ask Questions Like a Scientist

Quick Code: us4947s

Firefly Light Show

Watch the video. Then, **answer** the questions that follow.

Video

Let's Investigate a Firefly Light Show

SEP **Asking Questions and Defining Problems**

A-E-I-O-U

Write 1–2 adjectives that describe what you saw in the video:

_____

Write an emotion that describes how a particular part of the video made you feel:

_____

Write about something you found interesting:

_____

_____

_____

Write about something that surprised you and made you think "Oh!":

_____

_____

_____

Write about something that made you think "Um...": something you would like to learn more about, related to communication among organisms:

_____

_____

_____

Activity 3

Observe Like a Scientist

Alphabet and Written Language

Quick Code:
us4948s

Watch the video without stopping or taking notes. **Look** for examples of how communications have changed from simple to more complex.

Alphabet and Written Language

Talk Together

Now, talk together about the different types of communication you saw in the video. What are some similarities and differences between ancient writing systems and today's alphabet?

Activity 4
Observe Like a Scientist

Flags

Use the image below to spell out your name using the semaphoric alphabet. For each letter of your name, **draw** a person or stick figure holding the flags in the correct position.

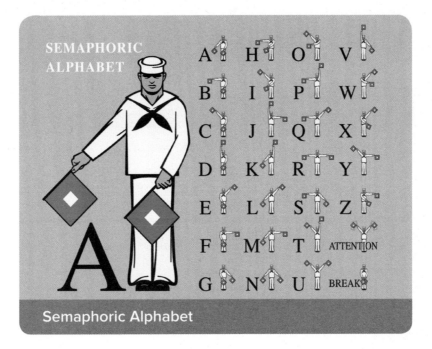

Semaphoric Alphabet

5.4 | Wonder

How do humans use light and other electromagnetic radiation to send and receive information?

Activity 5

Evaluate Like a Scientist

What Do You Already Know About Communication and Information Transfer?

Quick Code:
us4950s

Analog and Digital

Read the list of ways people communicate. **Classify** each type of communication in the table as analog (A) or digital (D).

Type of Communication	Is It Analog (A) or Digital (D)?
Talking in person	
Computers	
Food labels	
Street signs	
Using a cell phone	
E-reader	

Communication Systems

Draw a line to match each communication system to the way it sends messages.

Communication System	How the System Sends Messages
Signals are sent as radio waves through the air and received on devices that turn them into sound.	television
Signals are sent as radio waves from a device to a tower, then to another device.	Internet
Signals are encoded as packets of data and sent through cables underground to devices that decode them.	radio
Signals are sent as radio waves through the air and received on devices that turn them into sound and pictures.	cell phones

Activity 6

Analyze Like a Scientist

Transferring Information

Quick Code:
us4951s

Read the text. As you read, **mark** anything you don't understand with a question mark "?" and anything you find interesting with an exclamation point "!".

Transferring Information

Your sense organs collect information about your environment and send it to your brain. Examples include your ears that detect sound energy and your eyes that use light energy to gather information. For a moment, think about all the different kinds of information that you receive through your eyes. Your eyes detect light. This means they can detect signals that travel very fast over different distances, such as your friend waving from across a room, a traffic signal, or a rescue flare. In the past, people used signal fires to communicate over distances of many kilometers. Many backcountry hikers carry mirrors that they can flash to attract the attention of rescue helicopters.

SEP **Constructing Explanations and Designing Solutions**

Traffic Signals

Humans use **codes** to transmit information. They can be as simple as a thumbs-up or thumbs-down, or a red or a green light. Expressions on our faces are coded signals that can help people predict what we are thinking or whether we feel happy or sad. Dogs are also experts at reading human faces. Language is a code in sound. Different languages are different codes but they all enable the transfer of information. Writing is a code that uses symbols. A code is a pattern that has meaning, such as the arrangement of letters in a word. In some societies, drums are used to send information encoded in different rhythms. Lighthouses encode information in flashes of light that tell sailors where they are. Patterns of smoke rising into the sky from a fire can be used to encode information. When your sense organs receive this information and send it to your brain, your brain must decode the information. If it has not learned the code, then it cannot decode the information.

Activity 7

Think Like a Scientist

Quick Code:
us4952s

Inventing a Code

In this investigation, you will invent a code that is similar to Morse code.

Read the directions and **complete** the activity below. Then, **answer** the questions that follow.

What Will You Do?

1. With your partner, decide whether you will use a flashlight or a drum to communicate.

2. Then, work with your partner to create a unique signal for every letter of the alphabet.

3. Each partner should write down the code in the space below.

© Discovery Education | www.discoveryeducation.com • Image: Hero Images / DigitalVision / Getty Images

CCC Patterns

Discovery EDUCATION

4. Now, work with your partner to design a procedure for sending and receiving signals. Be sure to ask your teacher to check the procedures before you move on.

5. Talk with your partner to decide who will act as the person sending the message and who will act as the person receiving the message. Then, follow the directions below for the role that you chose.

If you are sending the message:

A. On a separate sheet of paper, write a unique message that is no more than five words. Then, use your code from step #3 to encode your message.

B. When your teacher instructs you to do so, stand across the room from your partner and use either the flashlight or the drum to send your encoded message to the receiver.

If you are receiving the message:

A. When your teacher instructs you to do so, stand across the room from your partner and wait to receive the message.

B. Then, use the space below to write down the coded message from the sender.

C. Now, use the code from step #3 to decode the message that you received.

D. Once the receiver has decoded the message, the receiver should talk with the sender to compare the message that was sent to the message that was received.

Think About the Activity

Did your message make it from your sender to your receiver correctly?
If not, what went wrong?

_____

_____

_____

_____

_____

_____

What sense did you use to receive your code ?

_____

_____

_____

_____

_____

_____

_____

What would you do to improve your code for future use?

_____

_____

_____

_____

_____

_____

Why are codes useful for humans who need to communicate across distances?

_____

_____

_____

_____

_____

_____

What Is the Difference between an Analog and Digital Signal?

Activity 8
Analyze Like a Scientist

Digital vs. Analog

Quick Code:
us4953s

Read the text. Then, **complete** the table and **answer** the questions that follow.

Digital vs. Analog

When you play a video game, what type of controller do you like to use? Your answer usually depends on the type of game you play. Let's say you play a game where you control an elf. You have to walk around a twisting and turning path. Now look at the two controllers below. Which one would you use? Instead, what if you were playing a game where you had to move

This directional controller is a type of digital controller.

a bug across a busy street one direction at a time—left, right, forward, or backward? Would you use the same controller? Probably not. The controller you choose depends on whether the information you share with the video game is **digital** or **analog**. So, what is the difference between digital and analog information? Digital information is shown by a number, or by a switch being turned off or on. Because a digit is a number, think of things that you can count when you think of digital. Analog information is continuous. Think of a dial, or a hand on a clock that constantly moves.

Turning a light on and off is like sending a digital signal.

Using a dimmer switch is like sending an analog signal.

You have many things around your home that can be described as either digital or analog. Things that turn on or off are digital. So are things that you push single buttons on, like keyboards. What analog items do you have? Look for knobs and sliders, like volume dials.

There are other types of information that you may not realize are digital. Think about the music you listen to. If you play a record from your grandparents' collection, that is an analog recording. The grooves on the record are a continuous recording of the actual sounds you hear. CDs, MP3 players, and other devices use digital recordings of music.

Digital vs. Analog *cont'd*

The sound waves are changed to numbers. When you are ready to play the music, the numbers are changed back to sound.

You have probably also seen analog and digital photos. Cameras that take analog photos use film. The film is developed and printed on paper. Digital cameras change the pictures you take into numbers. The numbers represent tiny dots of color. Then, computer monitors and printers change the numbers back to pictures again.

Digital and analog both have advantages and disadvantages. Analog data is thought to be purer data. For example, sound recordings are more accurate. Photos can have more details. Analog data takes up a lot of space. It also degrades, or breaks down, over time. Records and tapes can decay and wear down. Film and photos can age and fade.

Digital data takes up very little space. It also does not degrade much over time. As long as a computer or other device can read the digits, the data will last. However, digital data does not capture all of the detail that analog does. It can also be changed easily. For example, you can use software to change digital photos very simply. Off-key singers can be changed to sound pitch perfect on digital recordings.

So, is one format better than the other? It depends on what you want to do with the data. If you want to carry a lot of music with you, digital takes up less space. If you want more detailed photos, you probably want to use analog. Start looking around your home to see where you use each type of data.

	Examples	Advantages	Disadvantages
Digital			
Analog			

How can you tell the difference between an analog clock and a digital clock?

_____

_____

_____

_____

_____

Which form of data takes up less space?

_____

_____

_____

Which form of data records details more accurately?

_____

_____

_____

What does it mean to degrade?

_____

_____

_____

Activity 9

Think Like a Scientist

Going Digital

Complete the activity. Then, **answer** the questions that follow.

What Will You Do?

To create the digital image, **draw** the image of the bicycle on the grid. Every black area from the bicycle image should be filled in on the grid and any square that is filled in must be filled in completely. Also, every white area on the bicycle image should be left blank on the grid.

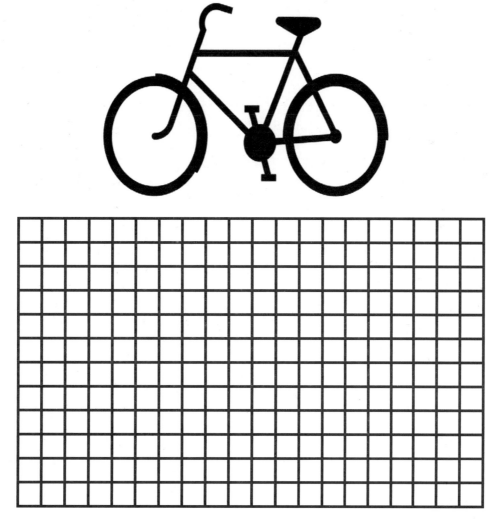

CCC **Patterns**

Think About the Activity

How is the digital image like the original? How is it different?

_____

_____

_____

_____

_____

How could you make the digital image more closely resemble the original?

_____

_____

_____

_____

How would doing so be different from your creation of this digital image?

_____

_____

_____

_____

Discovery
EDUCATION

Activity 10

Analyze Like a Scientist

Communication Systems

Quick Code:
us4955s

Read the text. Then, **underline** the parts of a communication system.

Communication Systems

When we use a cell phone, a computer connected to the Internet, or cable TV, we are using communication **systems**. These phone systems, Internet systems, and cable TV systems are all systems that communicate using signals. Each system consists of many parts that work together to transfer information from one place to another. Systems exist in nature or are designed by humans. Recall that your nervous system consists of many parts—nerves, brain, and sense organs that must work together to transfer information to and around the body. Similarly, a cell phone by itself cannot help you talk to your friends. It needs to be part of a system with other parts such as satellites, communication towers, and software. The parts of a system interact. This is why when all the parts are working properly, a system can perform in a way that the individual parts cannot.

Satellite Dishes and Cell Phone Towers

CCC Systems and System Models

Activity 11

Observe Like a Scientist

How Fiber Optics Work

Watch the video without stopping or taking notes. **Look** for examples of how fiber-optic cables work in a communication system.

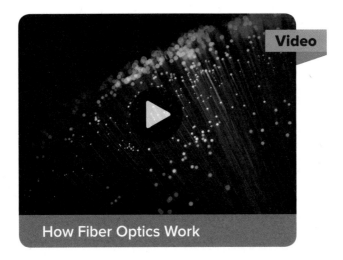

How Fiber Optics Work

Talk Together

Now, use your new ideas to discuss how fiber-optic cables are similar to the nervous system.

Activity 12

Observe Like a Scientist

Cell Phones

Watch the video without stopping or taking notes. **Look** for information about the different parts of a cell phone system.

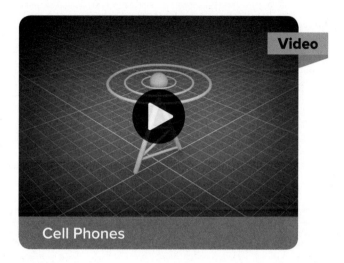

Cell Phones

Draw a diagram or flowchart showing the different parts of a cell phone system. Use arrows to show how information is transferred.

CCC **Systems and System Models**

Activity 13
Evaluate Like a Scientist

Quick Code:
us4958s

Cell Phone System

Each box represents a different part of a cell phone's communication system. **Use** the table to complete the correct order in which the parts work to send information from you to a friend when you talk on the phone.

| A. antenna receives radio waves | B. radio tower transmits radio waves | C. transmitter sends signals as radio waves | D. speaker changes radio waves into sound waves | E. microphone changes sound waves to electrical signals |

How a Cell Phone Sends Information	Parts of a Cell Phone's Communication System
First part	
Second part	
Third part	
Fourth part	
Fifth part	

SEP   Developing and Using Models

CCC   Systems and System Models

Design an Information Transfer System

The word bank contains is a list of commonly used circuit-building materials. **Think** about how you could use materials from the list to design an information transfer system. Then, **draw** a model of your system in the box.

Common Circuit-Building Materials

Batteries	Light bulbs	Solar panels
Wires	Buzzers	Light detectors
Switches	Speakers	Microphones

Activity 14

Record Evidence Like a Scientist

Quick Code:
us4959s

Firefly Light Show

Now that you've learned about communication and information transfer, watch the video Firefly Light Show again. You first saw this in Wonder.

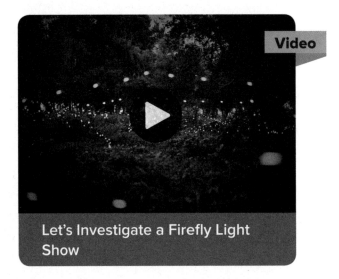

Let's Investigate a Firefly Light Show

Talk Together

How can you describe a firefly light show now?

How is your explanation different from before?

SEP Constructing Explanations and Designing Solutions

Look at the Can You Explain? question. You first read this at the beginning of the lesson.

Can You Explain?

How do humans use light and other electromagnetic radiation to send and receive information?

Now you will use your new ideas about the firefly light show to answer a question.

1. **Choose** a question. You can use the Can You Explain? question or one of your own. You can also use one of the questions that you wrote at the beginning of the lesson.

My Question

2. Then, **use** the graphic organizers on the next pages to help you **answer** the question.

To plan your scientific explanation, first **write** your claim. Your claim is a one-sentence answer to the question you investigated. It answers: What can you conclude? It should not start with yes or no.

My claim:

_____

_____

_____

_____

Finally, **explain** your reasoning. Reasoning ties together the claim and the evidence. Reasoning shows how or why the data count as evidence to support the claim. It may be helpful to color code your pieces of evidence and which portion of the explanation the evidence supports.

Evidence	How It Supports Claim

Now, **write** your scientific explanation.

_____

_____

_____

_____

_____

_____

_____

_____

_____

_____

_____

_____

_____

_____

_____

_____

_____

_____

STEM in Action

Activity 15

Analyze Like a Scientist

Cell Phones

What do you like to use a cell phone for? Do you play games on it? Or make calls? Or look on the Internet? As you know, cell phones can be used for a great many things, especially when they are connected to the Internet. But how can a cell phone connect to the Internet or make calls without a cable? How is the signal processed? In other words, how do cell phones work? **Read** the text to find out. Then, **complete** the activities that follow.

How a Cell Phone Works

It seems everyone has a cell phone that can fit in a pocket and travel anywhere. Yet, it wasn't that long ago that all telephones were big, bulky, and connected to wires. You certainly couldn't take them on a trip. You couldn't even pick a telephone up and take it into the next room.

That was then. Today's telephones fit in your hand. You can take them anywhere, and most people do. While it's true that you can use these devices to talk to your friends, they can also do so much more. They can take pictures and can store information, including photos and videos. They can hook up to the Internet. They have built-in calculators, and they can receive emails.

CCC **Systems and System Models**

Many people use radios to listen to the music, ball games, and the news. It might seem that cell phones are very different from radios, but they actually work in a similar way. Both allow us to communicate using radio signals that travel through the air. When you speak into a cell phone, the device takes your voice and converts it into a digital electrical signal. The phone then transmits the digital signal as radio waves through the air. Radio waves are a form of electromagnetic radiation, or light.

Cell phones send and receive all types of information over radio waves.

When someone calls you or sends you a picture, your cell phone acts as a receiver. It receives the digital radio signal and converts it back into an electrical signal and then sounds and pictures.

Cell phones don't have a lot of power. They can only send a signal over short distances. Still, you're able to talk to your aunt who lives clear across the country. Why is that? In order for your aunt, or anyone else to hear you, cell phone signals are relayed over a network that is divided into specific areas called "cells." Each cell in the network has a special station designed to send and receive the radio signals coming from your phone. Each station has a cell phone tower, which is nothing more than a radio **antenna**. These antennas relay your cell phone signal to another cell phone tower near the person you're calling. If you're moving in a car, your cell phone switches antennas as you move into different cells so that your call is not interrupted.

How are cell phones and radios similar?

_____

_____

_____

_____

What is one main difference between cell phones and radios?

_____

_____

_____

_____

_____

_____

_____

Why do you think so many people have a cell phone?

_____

_____

_____

What does the word *relay* mean in this passage?

_____

_____

_____

Antenna Placement

As you just read, antennas are an essential part of the external technology that makes cell phones work. Where do you think they should be placed? What would you need to consider to determine where you would place them? How do they make them blend into the landscape? How do cell phone companies decide where to locate their antennas?

There is not one set person who makes decisions about where to install cell phone towers. However, you will always find a cell tower installer leading the effort. The cell tower installer is responsible for a wide range of tasks. This person must know how cell technology works. He or she must understand how to repair and install high-tech equipment. The installer must have an excellent safety record and must be able to work with others.

Cell Phone Tower

Imagine that you are tasked with setting up a cell phone network in a new area. This means that you have to decide on the best locations to place two new cell phone towers.

The best places to put towers are on high points because the signal can usually travel for greater distances. Cell phone installers do not walk across the land and measure elevation—they use maps! A topographical map shows the elevation of different regions. A line on the map represents an elevation. The closer together the lines are, the steeper the land. A closed circle represents a peak or a high point.

Study the map below and **draw** an "X" to represent cell phone towers on areas where you think they will work best. You must draw two cell phone towers. To get the best service, do not draw your towers too close together.

Topographic Map

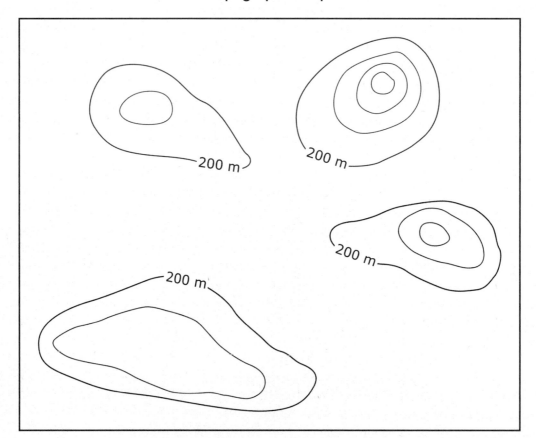

Explain the Answer

Explain why you placed the towers where you did. What factors did you take into consideration?

_____

_____

_____

_____

_____

_____

_____

_____

_____

_____

_____

_____

5.4 | Share

How do humans use light and other electromagnetic radiation to send and receive information?

Activity 16

Evaluate Like a Scientist

Quick Code:
us4961s

Review:
Communication and Information Transfer

Think about what you have read and seen in this lesson. **Write** down some core ideas you have learned. **Review** your notes with a partner. Your teacher may also have you take a practice test.

_____

_____

_____

_____

_____

_____

_____

_____

_____

_____

_____

_____

_____

SEP Obtaining, Evaluating, and Communicating Information

Talk Together

Think about what you saw in Get Started. Use your new ideas to discuss what systems are made of, how parts of a system work together, and how humans and other living things use communication systems.

Solve Problems Like a Scientist

Quick Code:
us4963s

Unit Project: Bat Chat

In this project, you will research bats to learn how their adaptations help them to navigate and communicate.

Read the text about echolocation. **Underline** the way bats use sound.

Chattering Bats

Many creatures use sound to communicate with each other. But sound can be used for other purposes. For example, bats use sound to communicate with each other. They also use sound to move around in the dark.

Bats live in dark places, such as caves. There is not enough light for them to see. Bats also fly very fast. They need to be able to avoid flying into walls and other objects. To do this, they have a special adaptation. They make a noise in their throats that is very high pitched. It is so high that humans cannot hear it. The noise bounces off objects, a process called echoing. Bats hear the echo with their ears. They use the echo to figure out where objects are. This way, they can avoid flying into objects. This is called echolocation.

SEP   Obtaining, Evaluating, and Communicating Information

CCC   Structure and Function

Bats also use echolocation to hunt. They make a noise, and the noise bounces off prey. Bats can find even tiny prey this way. For example, many bats eat mosquitoes. Although mosquitoes are very small, bats can find them with sound.

Bats also communicate with each other using sound. Bats make different sounds that mean different things, just like people communicate with words. Just as the sound "no" is different from "yes" for people, some sounds mean different things than others for bats.

Bats talk a lot. Most of the sounds are too high for humans to hear. Researchers use recording devices that can measure the sound. They have decoded many of the sounds bats make and have found that most of the sounds are arguments. Bats argue almost constantly. They argue about food. They argue about where they get to sleep. They argue about which bats they get to have as mates.

Bat Chat

Echolocation

Research bats further by using print or online sources. **Learn** about the ways bats have adapted to use sound to navigate, hunt, and communicate. Then, **draw** a diagram of a bat using sound to avoid obstacles and find prey. **Label** all relevant parts of the diagram. Be sure to include the way the sound interacts with the bat, the obstacles, and the prey.

Bat Chat

Bats communicate by using different sounds to mean different things, like humans use language. Bats also hunt and fly in the caves where they live, and they do so using echolocation.

Explain why it is helpful for bats to have different sounds that mean different things, given these facts. **Use** a Claim-Evidence-Reasoning chart to organize your thoughts.

Claim:	
Evidence	**Reason**

Grade 4 Resources

- **Bubble Map**
- **Safety in the Science Classroom**
- **Vocabulary Flash Cards**
- **Glossary**
- **Index**

Name _____

Bubble Map

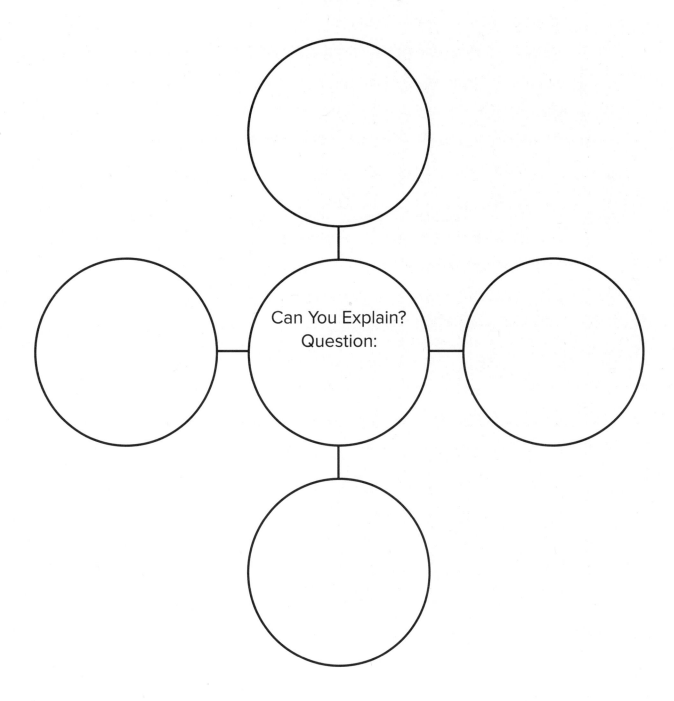

Can You Explain?
Question:

Safety in the Science Classroom

Following common safety practices is the first rule of any laboratory or field scientific investigation.

Dress for Safety
One of the most important steps in a safe investigation is dressing appropriately.

- Splash goggles need to be kept on during the entire investigation.

- Use gloves to protect your hands when handling chemicals or organisms.

- Tie back long hair to prevent it from coming in contact with chemicals or a heat source.

- Wear proper clothing and clothing protection. Roll up long sleeves, and if they are available, wear a lab coat or apron over your clothes. Always wear close toed shoes. During field investigations, wear long pants and long sleeves.

Be Prepared for Accidents
Even if you are practicing safe behavior during an investigation, accidents can happen. Learn the emergency equipment location in your classroom and how to use it.

- The eye and face wash station can help if a harmful substance or foreign object gets into your eyes or onto your face.

- Fire blankets and fire extinguishers can be used to smother and put out fires in the laboratory. Talk to your teacher about fire safety in the lab. He or she may not want you to directly handle the fire blanket and fire extinguisher. However, you should still know where these items are in case the teacher asks you to retrieve them.

- Most importantly, when an accident occurs, immediately alert your teacher and classmates. Do not try to keep the accident a secret or respond to it by yourself. Your teacher and classmates can help you.

DISCOVERY
EDUCATION

Practice Safe Behavior

There are many ways to stay safe during a scientific investigation. You should always use safe and appropriate behavior before, during, and after your investigation.

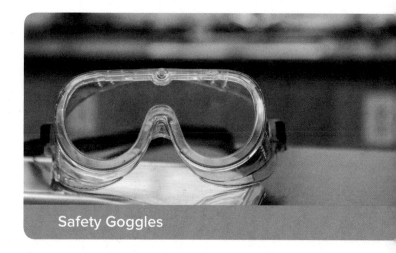
Safety Goggles

- Read the all of the steps of the procedure before beginning your investigation. Make sure you understand all the steps. Ask your teacher for help if you do not understand any part of the procedure.

- Gather all your materials and keep your workstation neat and organized. Label any chemicals you are using.

- During the investigation, be sure to follow the steps of the procedure exactly. Use only directions and materials that have been approved by your teacher.

- Eating and drinking are not allowed during an investigation. If asked to observe the odor of a substance, do so using the correct procedure known as wafting, in which you cup your hand over the container holding the substance and gently wave enough air toward your face to make sense of the smell.

- When performing investigations, stay focused on the steps of the procedure and your behavior during the investigation. During investigations, there are many materials and equipment that can cause injuries.

- Treat animals and plants with respect during an investigation.

- After the investigation is over, appropriately dispose of any chemicals or other materials that you have used. Ask your teacher if you are unsure of how to dispose of anything.

- Make sure that you have returned any extra materials and pieces of equipment to the correct storage space.

- Leave your workstation clean and neat. Wash your hands thoroughly.

adaptation

Image: Paul Fuqua

something a plant or animal does to help it survive in its environment

air

Image: Discovery Communications, Inc.

the part of the atmosphere closest to Earth; the part of the atmosphere that organisms on Earth use for respiration

analog

Image: Discovery Communications, Inc.

one continuous signal that does not have any breaks

antenna

Image: olafpictures/Pixabay

a device that receives radio waves and television signals

Arctic

Image: Discovery Communications, Inc.

being from an icy climate, such as the North Pole

brain

Image: Paul Fuqua

the main control center in an animal body; part of the central nervous system

camouflage

Image: Discovery Communications, Inc.

the coloring or patterns on an animal's body that allow it to blend in with its environment

code

Image: Pixabay

a way to communicate by sending messages using dots and dashes

digestive system

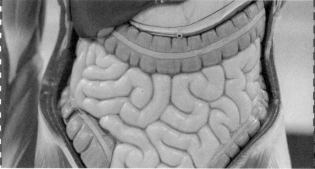

Image: Rattiya Thongdumhyu/Shutterstock

the body system that breaks down food into tiny pieces so that the body's cells can use it for energy

digital

Image: Discovery Communications, Inc.

a signal that is not continuous and is made up of tiny separate pieces

disease

Image: Discovery Communications, Inc.

a condition that disrupts processes in the body

ecosystem

Image: Paul Fuqua

all the living and nonliving things in an area that interact with each other

ear

Image: uaurelijus/Shutterstock

organ for hearing

electromagnetic spectrum

Image: Dr. Manuel Gonzales Reyes/Pixabay

the full range of frequencies of electromagnetic waves

energy

Image: Paul Fuqua

the ability to do work or cause change; the ability to move an object some distance

environment

Image: Odua Images / Shutterstock.com

all the living and nonliving things that surround an organism

extinct

Image: Ocean First Education

describes a species of animals that once lived on Earth but which no longer exists

heart

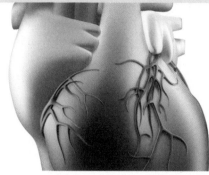

Image: decade3d - anatomy online / Shutterstock.com

the muscular organ of an animal that pumps blood throughout the body

heat

Image: Paul Fuqua

the transfer of thermal energy

hibernate

Image: Discovery Communications, Inc.

to reduce body movement during the winter in an effort to conserve energy

information

Image: Free-Photos/Pixabay

facts or data about something; the arrangement or sequence of facts or data

light

waves of electromagnetic energy; electromagnetic energy that people can see

Image: Paul Fuqua

matter

Image: Paul Fuqua

material that has mass and takes up some amount of space

migration

Image: Discovery Communications, Inc.

the movement of a group of organisms from one place to another, usually due to a change in seasons

nerve

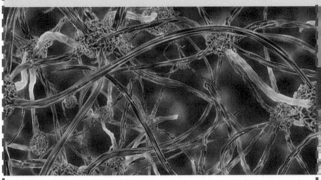

Image: Pixabay

a cell of the nervous system that carries signals to the body from the brain, and from the body to the brain and/or spinal cord

ocean

Image: Martin Winkler/Pixabay

a large body of salt water that covers most of Earth

opaque

Image: PavelShynkarou / Shutterstock.com

describes an object that light cannot travel through

organ

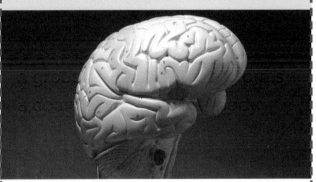

Image: Paul Fuqua

a group of tissues that performs a complex function in a body

organism

Image: Discovery Communications, Inc.

any individual living thing

pollute

Image: Paul Fuqua

to put harmful materials into the air, water, or soil

predator

Image: Ocean First Education

an animal that hunts and eats another animal

prey

Image: Discovery Communications, Inc.

an animal that is hunted and eaten by another animal

pupil

Image: Pixabay

the black circle at the center of an iris that controls how much light enters the eye

receptor

Image: JacLou DL/Pixabay

nerves located in different parts of the body that are especially adapted to receive information from the environment

reflect

Image: Discovery Communications, Inc.

light bouncing off of a surface

reflex

Image: PORTRAIT IMAGES ASIA BY NONWARIT/Shutterstock

an automatic response

refract

Image: Charlie Blacker/Shutterstock

to bend light as it passes
through a material

reproduce

Image: Paul Fuqua

to make more of a species; to
have offspring

seismic

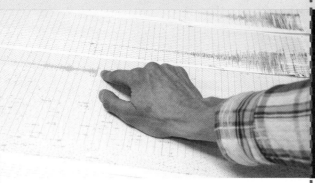

Image: Belish/Shutterstock

having to do with earthquakes or
earth vibrations

senses

Image: Nahid Sheikh/Pixabay

taste, touch, sight, smell, and
hearing

skin

Image: Discovery Communications, Inc.

an organ that covers and protects the bodies of many animals

sound

Image: Paul Fuqua

a vibration that travels through a material, such as air or water; something that you sense with your hearing

stimulus

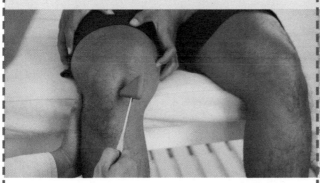

Image: Dragon Images/Shutterstock

things in the environment that cause us to react or have a physical response

stomach

Image: mrs. pandora/Pixabay

a muscular organ in the body where chemical and mechanical digestion take place

survive

Image: Paul Fuqua

to continue living or existing: An organism survives until it dies; a species survives until it becomes extinct.

system

Image: Gil Meshulam/Shutterstock

a group of related objects that work together to perform a function

thermal energy

Image: Paul Fuqua

energy in the form of heat

tongue

Image: Free-Photos/Pixabay

an organ in the mouth that helps in eating and speaking

trait

Image: Paul Fuqua

a characteristic or property of an organism

transparent

Image: Bernd Hildebrandt/Pixabayy

describes materials through which light can travel; materials that can be seen through

English ———— A ———— Español

acceleration
to increase speed

aceleración
aumentar la rapidez

adaptation
something a plant or animal does to help it survive in its environment (related word: adapt)

adaptación
algo que hace una planta o un animal para sobrevivir en su medio ambiente (palabra relacionada: adaptar)

air
the part of the atmosphere closest to Earth; the part of the atmosphere that organisms on Earth use for respiration

aire
parte de la atmósfera más cercana a la Tierra; la parte de la atmósfera que los organismos que habitan la Tierra utilizan para respirar

amplitude
height or "strength" of a wave

amplitud
altura o "magnitud" de una onda

analog
one continuous signal that does not have any breaks

analógico
señal continua que no tiene ninguna interrupción

antenna
a device that receives radio waves and television signals

antena
dispositivo que recibe ondas de radio y señales de televisión

Arctic
being from an icy climate, such as the North Pole

ártico
que pertenece a un clima helado, como el del polo norte

B

behavior
all of the actions and reactions of an animal or a person (related word: behave)

conducta
todas las acciones y reacciones de un animal o una persona (palabra relacionada: comportarse)

brain
the main control center in an animal body; part of the central nervous system

cerebro
principal centro de control en el cuerpo de un animal; parte del sistema nervioso central

C

camouflage
the coloring or patterns on an animal's body that allow it to blend in with its environment

camuflaje
color o patrones del cuerpo de un animal que le permiten confundirse con su medio ambiente

canyon

a deep valley carved by flowing water

cañón

valle profundo labrado por el flujo del agua

chemical energy

energy that can be changed into motion and heat

energía química

energía que se puede cambiar a movimiento y calor

chemical weathering

changes to rocks and minerals on Earth's surface that are caused by chemical reactions

meteorización química

cambios en las rocas y minerales de la superficie de la Tierra causados por reacciones químicas

code

a way to communicate by sending messages using dots and dashes

código

forma de comunicarse enviando mensajes con puntos y rayas

collision

the moment where two objects hit or make contact in a forceful way

colisión

el momento en el que dos objetos chocan o hacen contacto de forma contundente

conduction

when energy moves directly from one object to another

conducción

cuando la energía pasa en forma directa de un objeto a otro

conservation of energy

energy cannot be created or destroyed, it can only be changed from one form to another, such as when electrical energy is changed into heat energy

conservación de la energía

la energía no se puede crear o destruir; solo se puede cambiar de una forma a otra, como cuando la energía eléctrica cambia a energía térmica

conserve

to protect something, or prevent the wasteful overuse of a resource

conservar

proteger algo o evitar el uso excesivo e ineficiente de un recurso

convert (v)

to change forms

convertir (v)

cambiar de forma

delta

a fan-shaped mass of mud and other sediment that forms where a river enters a large body of water

delta

masa de barro y otros sedimentos parecida a un abanico, que se forma donde un río ingresa a un gran cuerpo de agua

deposition

laying sediment back down after erosion moves it around

sedimentación

volver a depositar sedimentos una vez que la erosión los arrastra

digestive system

the body system that breaks down food into tiny pieces so that the body's cells can use it for energy

sistema digestivo

sistema del cuerpo que descompone alimentos en pequeños trozos para que las células del cuerpo puedan usarlos para obtener energía.

digital

a signal that is not continuous and is made up of tiny separate pieces

digital

una señal que no es continua y está compuesta por diminutas partes separadas

disease

a condition that disrupts processes in the body and usually causes an illness

enfermedad

afección que perturba los procesos del cuerpo y por lo general produce una enfermedad

dune

a hill of sand created by the wind

duna

colina de arena creada por el viento

— **E** —

ear

organ for hearing

oído

órgano para oir

Earth

the third planet from the sun; the planet on which we live (related words: earthly; earth – meaning soil or dirt)

Tierra

tercer planeta desde el Sol; planeta en el cual vivimos (palabras relacionadas: terrenal; tierra en el sentido de suelo o suciedad)

earthquake

a sudden shaking of the ground caused by the movement of rock underground

terremoto

repentina sacudida de la tierra causada por el movimiento de rocas subterráneas

ecosystem

all the living and nonliving things in an area that interact with each other

ecosistema

todos los seres vivos y objetos sin vida de un área, que se interrelacionan entre sí

electromagnetic spectrum

the full range of frequencies of electromagnetic waves

espectro electromagnético

rango completo de frecuencias de las ondas electromagnéticas

energy

the ability to do work or cause change; the ability to move an object some distance

energía

capacidad de hacer un trabajo o producir un cambio; capacidad de mover un objeto a cierta distancia

energy source

where a form of energy begins

fuente de energía

origen de una forma de energía

energy transfer

the transfer of energy from one organism to another through a food chain or web; or the transfer of energy from one object to another, such as heat energy

transferencia de energía

transmisión de energía de un organismo a otro a través de una cadena o red alimentaria; o transmisión de energía de un objeto a otro, como por ejemplo la energía calórica

engineer

Engineers have special skills. They design things that help solve problems.

ingeniero

los ingenieros poseen habilidades especiales. Diseñan cosas que ayudan a resolver problemas.

environment

all the living and nonliving things that surround an organism

medio ambiente

todos los seres vivos y objetos sin vida que rodean a un organismo

erosion

the removal of weathered rock material. After rocks have been broken down, the small particles are transported to other locations by wind, water, ice, and gravity.

erosión

eliminación de material rocoso desgastado. Después de descomponerse las rocas, el viento, el agua, el hielo y la gravedad transportan las partículas pequeñas a otros lugares

erupt

the action of lava coming out of a hole or crack in Earth's surface; the sudden release of hot gasses or lava built up inside a volcano (related word: eruption)

erupción

acción de la lava que sale de un agujero o cráter de la superficie de la Tierra; repentina liberación de gases calientes o lava que se acumulan en el interior de un volcán (palabra relacionada: erupción)

extinct

describes a species of animals that once lived on Earth but which no longer exists (related word: extinction)

extinto

palabra que hace referencia a una especie de animales que habitaba antiguamente la Tierra, pero que ya no existe (palabra relacionada: extinción)

F

fault

a fracture, or a break, in the Earth's crust (related word: faulting)

falla

fractura, o quiebre, en la corteza de la Tierra (palabra relacionada: fallas)

feature

things that describe what something looks like

rasgo

cosas que describen cómo se ve algo

force

a pull or push that is applied to an object

fuerza

acción de atraer o empujar que se aplica a un objeto

forecast

(v) to analyze weather data and make an educated guess about weather in the future; (n) a prediction about what the weather will be like in the future based on weather data

pronosticar / pronóstico

(v) analizar los datos del tiempo y hacer una conjetura informada sobre el tiempo en el futuro; (s) predicción sobre cómo será el tiempo en el futuro con base en datos

fossil fuels

fuels that come from very old life forms that decomposed over a long period of time, like coal, oil, and natural gas

combustibles fósiles

combustibles que provienen de formas de vida muy antiguas que se descompusieron en el transcurso de un período de tiempo largo, como el carbón, el petróleo y el gas natural

friction

a force that slows down or stops motion

fricción

fuerza que desacelera o detiene el movimiento

fuel

any material that can be used for energy

combustible

todo material que puede usarse para producir energía

--- G ---

generate

to produce by turning a form of energy into electricity

generar

producir convirtiendo una forma de energía en electricidad

geothermal

heat found deep within Earth

geotérmica

calor que se encuentra en la profundidad de la Tierra

glacier

a large sheet of ice or snow that moves slowly over Earth's surface

glaciar

gran capa de hielo o nieve que se mueve lentamente sobre la superficie de la Tierra

gravitational potential energy
energy stored in an object based on its height and mass

energía potencial gravitacional
energía almacenada en un objeto que depende de su altura y masa

gravity
the force that pulls an object toward the center of Earth (related word: gravitational)

gravedad
fuerza que empuja a un objeto hacia el centro de la Tierra (palabra relacionada: gravitacional)

 H

heart
the muscular organ of an animal that pumps blood throughout the body

corazón
órgano muscular de un animal que bombea sangre a través del cuerpo

heat
the transfer of thermal energy

calor
transferencia de energía térmica

hibernate

to reduce body movement during the winter in an effort to conserve energy (related word: hibernation)

hibernar

reducir el movimiento del cuerpo durante el invierno con la finalidad de conservar la energía (palabra relacionada: hibernación)

I

---

information

facts or data about something; the arrangement or sequence of facts or data

información

hechos o datos sobre algo; la organización o secuencia de hechos o datos

---

K

---

kinetic energy

the energy an object has because of its motion

energía cinética

energía que posee un objeto a causa de su movimiento

L

landform

a large natural structure on Earth's surface, such as a mountain, a plain, or a valley

accidente geográfico

estructura natural grande que se encuentra en la superficie de la Tierra, como una montaña, una llanura o un valle

lava

molten rock that comes through holes or cracks in Earth's crust that may be a mixture of liquid and gas but will turn into solid rock once cooled

lava

roca fundida que sale por orificios o grietas en la corteza terrestre, y que puede ser una mezcla de líquido y gas pero se convierte en roca sólida al enfriarse

light

a form of energy that moves in waves and particles and can be seen

luz

forma de energía que se desplaza en ondas y partículas y que puede verse

magma

melted rock located beneath Earth's surface

magma

roca fundida que se encuentra debajo de la superficie de la Tierra

magnetic field

a region in space near a magnet or electric current in which magnetic forces can be detected

campo magnético

región en el espacio cerca de un imán o de una corriente eléctrica, donde pueden detectarse fuerzas magnéticas

map

a flat model of an area

mapa

modelo plano de un área

mass

the amount of matter in an object

masa

cantidad de materia que hay en un objeto

matter

material that has mass and takes up some amount of space

materia

material que tiene masa y ocupa cierta cantidad de espacio

meander

winding or indirect movement or course

meandro

movimiento o curso serpenteante o indirecto

migration

the movement of a group of organisms from one place to another, usually due to a change in seasons

migración

desplazamiento de un grupo de organismos de un lugar a otro, generalmente debido a un cambio de estaciones

model

a drawing, object, or idea that represents a real event, object, or process

modelo

dibujo, objeto o idea que representa un suceso, objeto, o proceso real

motion

when something moves from one place to another (related words: move, movement)

movimiento

cuando algo pasa de un lugar a otro (palabra relacionada: mover, desplazamiento)

mountain

an area of land that forms a peak at a high elevation (related term: mountain range)

montaña

área de tierra que forma un pico a una elevación alta (palabra relacionada: cadena montañosa)

nerve

a cell of the nervous system that carries signals to the body from the brain, and from the body to the brain and/or spinal cord

nervio

célula del sistema nervioso que lleva señales al cuerpo desde el cerebro, y desde el cuerpo al cerebro y/o médula espinal

nonrenewable

once it is used, it cannot be made or reused again

no renovable

una vez usado, no puede rehacerse o reutlizarse

nonrenewable resource

a natural resource of which a finite amount exists, or one that cannot be replaced with currently available technologies

recurso no renovable

recurso natural del cual existe una cantidad finita, o uno que no puede remplazarse con las tecnologías actualmente disponibles

nuclear energy

the energy released when the nucleus of an atom is split apart or combined with another nucleus

energía nuclear

energía liberada cuando el núcleo de un átomo se divide o combina con otro núcleo

ocean
a large body of salt water that covers most of Earth

océano
gran cuerpo de agua salada que cubre la mayor parte de la Tierra

opaque
describes an object that light cannot travel through

opaco
describe un objeto que la luz no puede atravesar

organ
a group of tissues that performs a complex function in a body

órgano
conjunto de tejidos que realizan una función compleja en el cuerpo

organism
any individual living thing

organismo
todo ser vivo individual

photosynthesis
the process in which plants and some other organisms use the energy in sunlight to make food

fotosíntesis
proceso por el cual las plantas y algunos otros organismos usan la energía del Sol para producir alimentos

pollute

to put harmful materials into the air, water, or soil (related words: pollution, pollutant)

contaminar

poner materiales perjudiciales en el aire, agua o suelo (palabras relacionadas: contaminación, contaminante)

pollution

when harmful materials have been put into the air, water, or soil (related word: pollute)

contaminación

cuando se introducen materiales perjudiciales en el aire, el agua o el suelo (palabra relacionada: contaminar)

potential energy

the amount of energy that is stored in an object; energy that an object has because of its position relative to other objects

energía potencial

cantidad de energía almacenada en un objeto; energía que tiene un objeto debido a su posición relativa con otros objetos

predator

an animal that hunts and eats another animal

depredador

animal que caza y come a otro animal

predict

to guess what will happen in the future (related word: prediction)

predecir

adivinar qué sucederá en el futuro (palabra relacionada: predicción)

prey

an animal that is hunted and eaten by another animal

presa

animal que es cazado y comido por otro

pupil

the black circle at the center of an iris that controls how much light enters the eye

pupila

círculo negro en el centro del iris que controla cuánta luz entra al ojo

--- R ---

radiant energy

energy that does not need matter to travel; light

energía radiante

energía que no necesita de la materia para desplazarse; luz

radiation

electromagnetic energy (related word: radiate)

radiación

energía electromagnética (palabra relacionada: irradiar)

receptor

nerves located in different parts of the body that are especially adapted to receive information from the environment

receptor

nervios ubicados en diferentes partes del cuerpo que están especialmente adaptados para recibir información del medio ambiente

---

reflect

light bouncing off a surface (related word: reflection)

reflejar

rebotar la luz sobre una superficie (palabra relacionada: reflexión)

---

reflex

an automatic response

reflejo

respuesta automática

---

refract

to bend light as it passes through a material (related word: refraction)

refractar

torcer luz cuando pasa a través de un material (palabra relacionada: refracción)

---

remote (adj)

to be operated from a distance

remoto (adj)

que se opera a distancia

renewable

to reuse or make new again

renovable

reutilizar o volver a hacer de nuevo

---

renewable resource

a natural resource that can be replaced

recurso renovable

recurso natural que puede reemplazarse

---

reproduce

to make more of a species; to have offspring (related word: reproduction)

reproducir

engendrar más individuos de una especie; tener descendencia (palabra relacionada: reproducción)

---

resistance

when materials do not let energy transfer through them

resistencia

cuando los materiales no permiten la transferencia de energía a través de ellos

---

resource

a naturally occurring material in or on Earth's crust or atmosphere of potential use to humans

recurso

material que se origina de forma natural en o sobre la corteza o la atmósfera de la Tierra, que es de uso potencial para los seres humanos

rotate

turning around on an axis;
spinning (related word: rotation)

rotar

girar sobre un eje; dar vueltas
(palabra relacionada: rotación)

— S —

satellite

a natural or artificial object that
revolves around another object
in space

satélite

objeto natural o artificial que gira
alrededor de otro objeto en el
espacio

sediment

solid material, moved by wind
and water, that settles on the
surface of land or the bottom of a
body of water

sedimento

material sólido que el viento o
el agua transportan y que se
asienta en la superficie de la
tierra o en el fondo de un cuerpo
de agua

seismic

having to do with earthquakes or
earth vibrations

sísmico

relativo a los terremotos o a las
vibraciones de la Tierra

seismic wave

waves of energy that travel
through the Earth

onda sísmica

ondas de energía que se
desplazan a través del interior
de la Tierra

senses

taste, touch, sight, smell, and
hearing (related word: sensory)

sentidos

gusto, tacto, visión, olfato y oído
(palabra relacionada: sensorial)

skin

an organ that covers and
protects the bodies of many
animals

piel

órgano que cubre y protege el
cuerpo de muchos animales

soil

the outer layer of Earth's crust
in which plants can grow; made
of bits of dead plant and animal
material as well as bits of rocks
and minerals

suelo

capa externa de la corteza de
la Tierra en donde crecen las
plantas; formada por pedazos de
plantas y animales muertos, así
como por pedazos de rocas y de
minerales

solar energy

energy that comes from the sun

energía solar

energia que proviene del Sol

sound

anything you can hear that travels by making vibrations in air, water, and solids

sonido

todo lo que se puede oír, que se desplaza produciendo vibraciones en el aire, el agua y los objetos sólidos

sound wave

a sound vibration as it is passing through a material: Most sound waves spread out in every direction from their source.

onda sonora

vibración que produce el sonido cuando atraviesa un material: la mayoría se dispersa desde la fuente en todas direcciones.

speed

the measurement of how fast an object is moving

rapidez

medida de la tasa a la que se desplaza un objeto

stimulus

things in the environment that cause us to react or have a physical response

estímulo

algo en el medio ambiente que nos hace reaccionar o tener una respuesta física

stomach

a muscular organ in the body where chemical and mechanical digestion take place

estómago

órgano muscular del cuerpo donde tiene lugar la digestión química y mecánica

---

sun

any star around which planets revolve

sol

toda estrella alrededor de la cual giran los planetas

---

survive

to continue living or existing: an organism survives until it dies; a species survives until it becomes extinct (related word: survival)

sobrevivir

continuar viviendo o existiendo: un organismo sobrevive hasta que muere; una especie sobrevive hasta que se extingue (palabra relacionada: supervivencia)

---

system

a group of related objects that work together to perform a function

sistema

grupo de objetos relacionados que funcionan juntos para realizar una función

tectonic plate
one of several huge pieces of
Earth's crust

placa tectónica
una de las muchas partes
enormes de la corteza terrestre

thermal energy
energy in the form of heat

energía térmica
energía en forma de calor

tongue
an organ in the mouth that helps
in eating and speaking

lengua
órgano de la boca que ayuda a
comer y hablar

topographic map
a map that shows the size and
location of an area's features
such as vegetation, roads, and
buildings

mapa topográfico
mapa que muestra el tamaño y la
ubicación de características de
un área, como la vegetación, las
carreteras y los edificios

trait
a characteristic or property of an
organism

rasgo
característica o propiedad de un
organismo

transparent

describes materials through which light can travel; materials that can be seen through

transparente

describe materiales a través de los cuales puede desplazarse la luz; materiales a través de los cuales se puede ver

tsunami

a giant ocean wave (related word: tidal wave)

tsunami

ola gigante en el océano (palabra relacionada: maremoto)

valley

a low area of land between two higher areas, often formed by water

valle

área baja de tierra entre dos áreas más altas, generalmente formada por el agua

volcano

an opening in Earth's surface through which magma and gases or only gases erupt (related word: volcanic)

volcán

abertura en la superficie de la Tierra a través de la cual surgen magma y gases, o solo gases, que hacen erupción (palabra relacionada: volcánico)

water

a compound made of hydrogen and oxygen; can be in either a liquid, ice, or vapor form and has no taste or smell

agua

compuesto formado por hidrógeno y oxígeno; puede estar en forma de líquido, hielo o vapor y no tiene sabor ni olor

wave

a disturbance caused by a vibration; waves travel away from the source that makes them

onda

perturbación causada por una vibración que se aleja de la fuente que la origina

wavelength

the distance between one peak and the next on a wave

longitud de onda

distancia entre un pico y otro en una onda

weathering

the physical or chemical breakdown of rocks and minerals into smaller pieces or aqueous solutions on Earth's surface

meteorización

desintegración física o química de rocas y minerales en partes más pequeñas o en soluciones acuosas en la superficie de la Tierra

work

a force applied to an object over a distance

trabajo

fuerza aplicada a un objeto a lo largo de una distancia

Index